THE HISTORY OF

SCIENCE

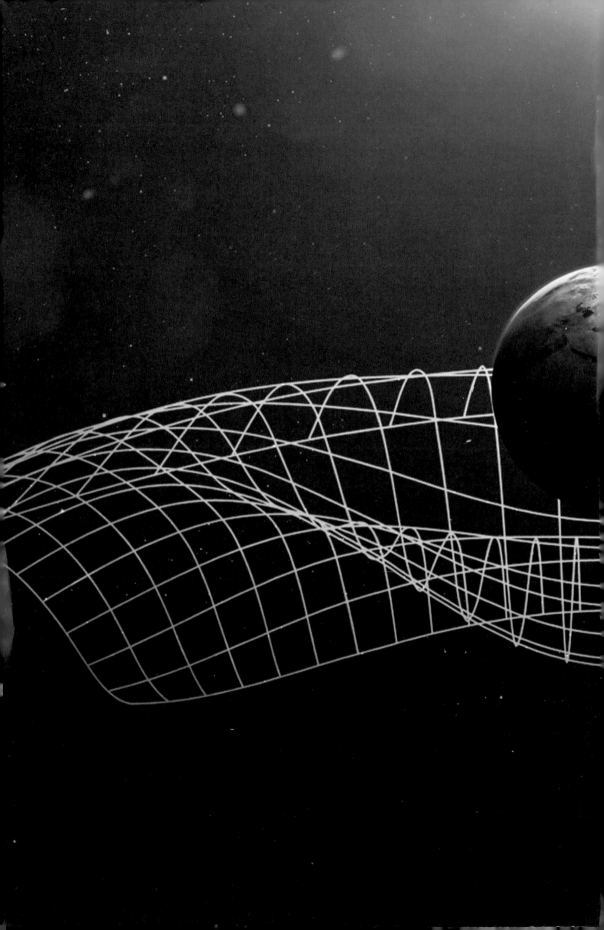

THE HISTORY OF
SCIENCE

TOM JACKSON

WORTH
WPRESS

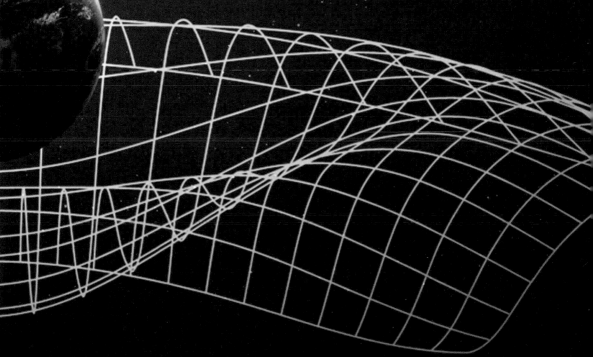

First published in 2019 by Worth Press Ltd, Bath, England
worthpress@btconnect.com

British Library Cataloguing in Publication Data
A catalogue record for this book is available from the British Library

ISBN: 978-1-84931-153-3

10 9 8 7 6 5 4 3 2 1

Publisher's Note Every effort has been made to ensure the accuracy of the information presented in this book. The publisher will not assume liability for damages caused by inaccuracies in the data and makes no warranty whatsoever expressed or implied. The publisher welcomes comments and corrections from readers, emailed to worthpress@btconnect.com, which will be considered for incorporation in future editions. Every effort has been made to trace copyright holders and seek permission to use illustrative and other material. The publisher wishes to apologize for any inadvertent errors or omissions and would be glad to rectify these in future editions.

The images used in this book come from either the public domain or from the public commons unless otherwise stated.

Design and layout: Chandra Creative & Content Solutions Pvt. Ltd.
Cover Design: Wayne Morrish

Printed and bound in China

HOW TO USE THE BOOK

Features section:
Provides information on various aspects of evolution and history of "science" as a branch of study.

Timeline section:
Entries on significant events flow down in chronological order (on blue background).

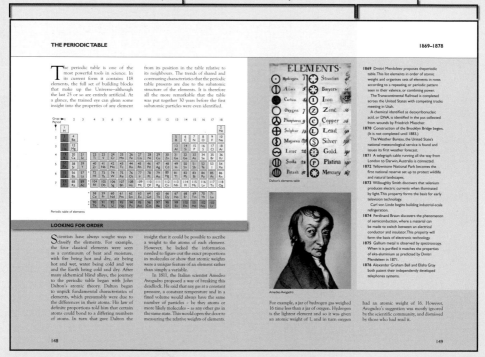

CONTENTS

The history of science is the best story ever told. It contains many great characters, it spans the ages and the action takes place in all corners of the globe – and even out in space – and perhaps most importantly of all it helps us to understand our place among all the rocks, oceans, animals, plants, planets, and stars, by revealing how the Universe works.

Science is an ancient idea, but we only really got to grip on it about 500 years ago. The word "science" comes from the Latin for "knowledge" and it began to be used in the 14th century to mean a skill or expertise. At this time, the people who wanted to investigate and understand nature were still known as natural philosophers, and they would retain that title up until the 1830s, when William Whewell, a clergyman who had a way with words, coined this term along with many others still used in science today.

Natural philosophers were a very mixed group. Some simply thought up answers to their questions. Clever as they may have been, these explanations did not stand up to scrutiny. One way to find out stuff was to test your ideas. A few natural philosophers took very significant steps forward using experiments, such as Eratosthenes who measured the Earth in the 3rd century BCE, Al-Haythem who pioneered optics in the 10th century CE, and Galileo Galilei who investigated the way bodies fall (among many other things) in the early 17th century. However, the process of doing science took shape in the mid 17th century. The scientific method holds true today and you could try it yourself: observe the system you are interested in and think about what you don't understand about it. Propose an explanation for your mystery – this is your hypothesis – and use it to predict the outcome of a test, or experiment. The result of the experiment should reveal if your hypothesis is correct or not.

In the 1930s, the Austrian philosopher Karl Popper turned this idea on its head somewhat. The experiment could only be conclusive about the falsity of a hypothesis. If the results did not show the idea was false then is was "not false", which was as close as science can get us to "true." Some might see that as a shortcoming, but it is really science's strength. No part of it is immune from being questioned.

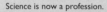
Science is now a profession.

The School of Athens, a painting by Raphael, shows all the great thinkers of ancient Greece.

An experiment can prove even the most established set of ideas to be false. In the 1960s, the American philosopher Thomas Kuhn described how this kind of "scientific crisis" was an essential part of pushing back the boundaries of knowledge. When the current paradigm, or collection of ideas and theories that underlie our understanding of the world, begins to accrue mysteries that cannot be explained, then science is in crisis. Something in the paradigm must be false, and once revealed, a new paradigm can take its place, a process that Kuhn called a paradigm shift. There have been some famous examples, such as Copernicus proving that Earth orbits the Sun, Charles's Darwin's theory of evolution, and Niels Bohr's quantum atomic model. All these form part of the plot, and so do the latest mysteries like dark energy and dark matter. Some scientists think we are in another scientific crisis today. Where will the paradigm shift next? Read on for the full story.

Karl Popper.

7

The Periodic Table is perhaps the most simple and effective chart in all of science. All of chemistry is on display, all the elements from the lightest to the heaviest are there at a glance. The table reveals much to the trained eye, but as a starting point the lightest elements are near the top, and the heaviest are at the bottom. The metals make

The periodic table of elements.

up the left side and the middle while the non-metals fill the right side. As well as a symbol, each element is shown with two numbers. The smaller one is the atomic number, which is the number of protons in the atoms of this element. The larger one is the mass number, which is a comparison of the relative weight of each atom.

									2 **He** Helium
				5 **B** Boron	6 **C** Carbon	7 **N** Nitrogen	8 **O** Oxygen	9 **F** Fluorine	10 **Ne** Neon
				13 **Al** Aluminum	14 **Si** Silicon	15 **P** Phosphorus	16 **S** Sulfur	17 **Cl** Chlorine	18 **Ar** Argon
27 **Co** Cobalt	28 **Ni** Nickel	29 **Cu** Copper	30 **Zn** Zinc	31 **Ga** Gallium	32 **Ge** Germanium	33 **As** Arsenic	34 **Se** Selenium	35 **Br** Bromine	36 **Kr** Krypton
45 **Rh** Rhodium	46 **Pd** Palladium	47 **Ag** Silver	48 **Cd** Cadmium	49 **In** Indium	50 **Sn** Tin	51 **Sb** Antimony	52 **Te** Tellurium	53 **I** Iodine	54 **Xe** Xenon
77 **Ir** Iridium	78 **Pt** Platinum	79 **Au** Gold	80 **Hg** Mercury	81 **Tl** Thallium	82 **Pb** Lead	83 **Bi** Bismuth	84 **Po** Polonium	85 **At** Astatine	86 **Rn** Radon
109 **Mt** Meitnerium	110 **Ds** Darmstadtium	111 **Rg** Roentgenium	112 **Cn** Copernicium	113 **Nh** Nihonium	114 **Fl** Flerovium	115 **Mc** Moscovium	116 **Lv** Livermorium	117 **Ts** Tennessine	118 **Og** Oganesson

62 **Sm** Samarium	63 **Eu** Europium	64 **Gd** Gadolinium	65 **Tb** Terbium	66 **Dy** Dysprosium	67 **Ho** Holmium	68 **Er** Erbium	69 **Tm** Thulium	70 **Yb** Ytterbium	71 **Lu** Lutetium
94 **Pu** Plutonium	95 **Am** Americium	96 **Cm** Curium	97 **Bk** Berkelium	98 **Cf** Californium	99 **Es** Einsteinium	100 **Fm** Fermium	101 **Md** Mendelevium	102 **Nd** Nobelium	103 **Lr** Lawrencium

We are still learning a lot about our neighbourhood in space. For most of recorded history the heavens had just the Moon, Sun and five planets from Mercury to Saturn. Astronomers have now mapped more than half a million bodies of one or the another kind from

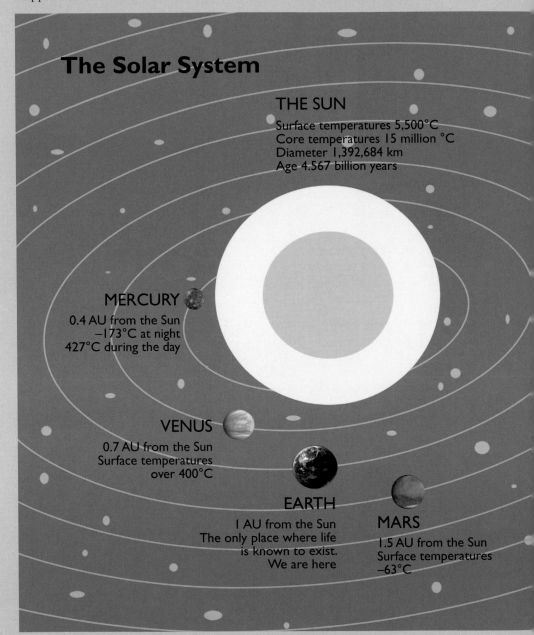

The Solar System

THE SUN
Surface temperatures 5,500°C
Core temperatures 15 million °C
Diameter 1,392,684 km
Age 4.567 billion years

MERCURY
0.4 AU from the Sun
−173°C at night
427°C during the day

VENUS
0.7 AU from the Sun
Surface temperatures
over 400°C

EARTH
1 AU from the Sun
The only place where life
is known to exist.
We are here

MARS
1.5 AU from the Sun
Surface temperatures
−63°C

the NEA, or Near Earth Asteroids, to the Kuiper Belt. There is much to investigate as you can see here, and you realize that there is more beyond that. A diffuse field of ice bodies called the Scattered Disc reacts to the Oort Cloud which surrounds our little patch of space.

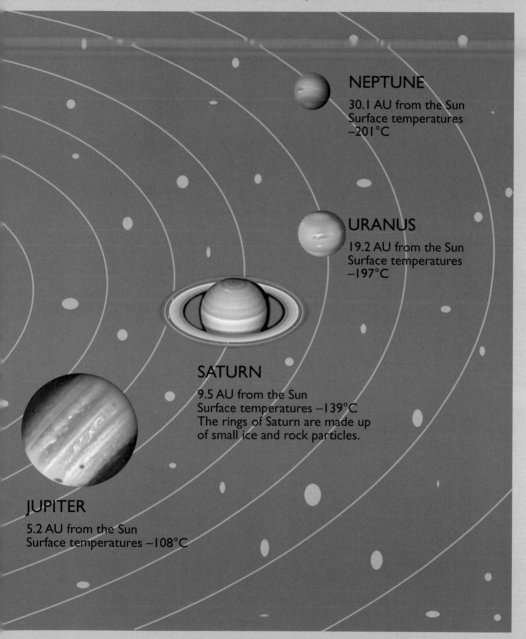

NEPTUNE

30.1 AU from the Sun
Surface temperatures
−201°C

URANUS

19.2 AU from the Sun
Surface temperatures
−197°C

SATURN

9.5 AU from the Sun
Surface temperatures −139°C
The rings of Saturn are made up
of small ice and rock particles.

JUPITER

5.2 AU from the Sun
Surface temperatures −108°C

Science has made it possible to look at a history of the Universe in one image. The Big Bang was named by the British scientist Fred Hoyle, who had hoped the term would help to dismiss the theory in favour of his own. However, science proved he was wrong and the Big Bang is still the biggest idea out there.

The Evolution of the Universe

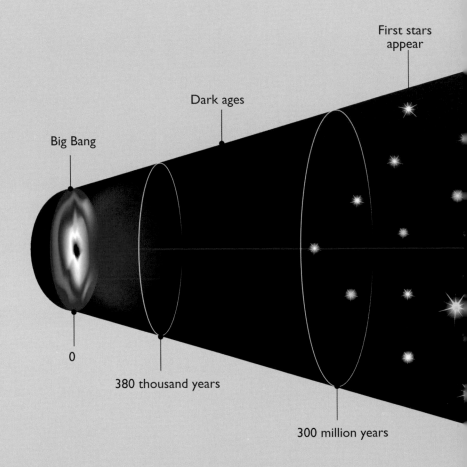

First stars
appear

Dark ages

Big Bang

0

380 thousand years

300 million years

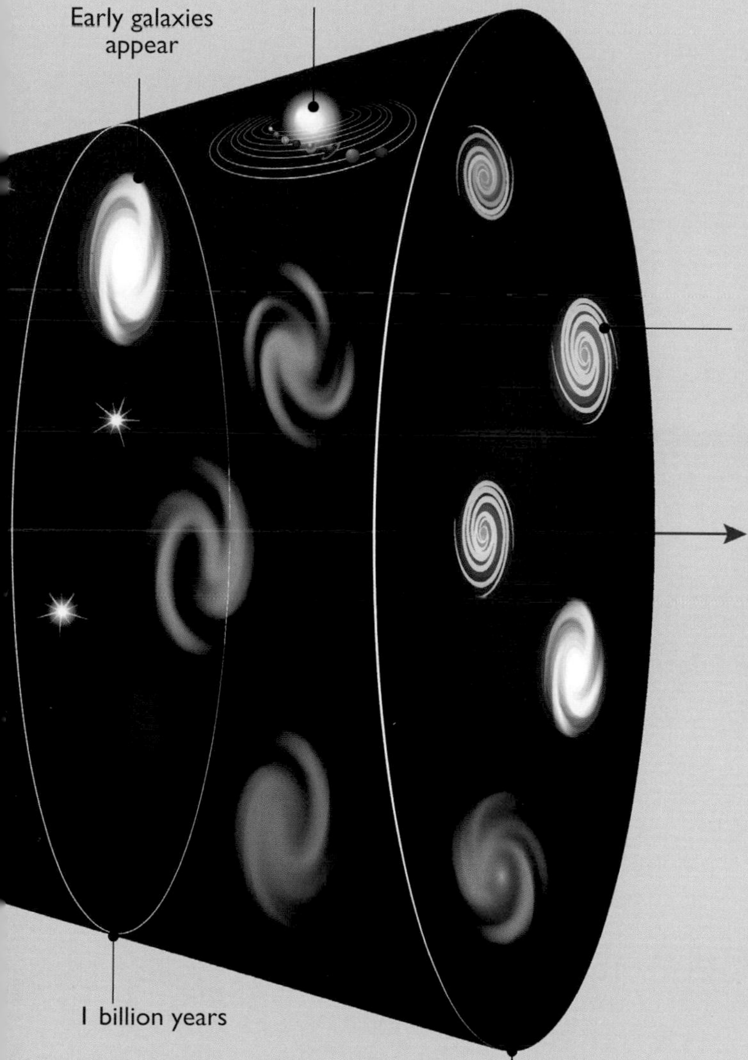

Formation of the
Solar System
(9 billion years)

Early galaxies
appear

Modern
galaxies

Time

1 billion years

Today

It is a sobering thought that science and technology are older than modern humans. Our distant prehuman ancestors, the Australopithecines, were making sense of the world and acting on those discoveries more than 3 million years ago. We know this from the stone tools they left behind that have been shaped into primitive but effective cutters and scrapers for butchering meat. There is little doubt that these early hominids were also fashioning bone, wood and shells too into useful implements, but these have mostly decayed in the interim. Only stone tools have survived the ages in any great numbers, hence our characterization of this part of history as the Stone Age.

Hand axes from Mauritania (center), Niger (left) and Israel (right).

Some more stone tools.

OLD STYLE

Of course, our Stone Age ancestors were not the only animals to make tools, but they were the first to use tools to make tools. The production of stone tools followed a widespread process that remained largely unchanged for the first 1.5 million years of the Stone Age. For much of this time, the hominid family tree was dominated by *Homo habilis*, or "handy man," so called because of his (and her) marked tool-making skills. The manufacturing technique required a rounded, hand-sized hard stone which was bashed on to a core stone, making it crack to reveal a sharp, wedge-like edge suitable for cutting. The hammer stone was often a rounded cobble taken from a stream bed, while the core stones were generally quartz-rich stones, such as chert and flint.

3.3 million years ago The earliest known direct ancestor of modern humans, Australopithecus, is making simple cutting and digging tools from stone, bone, and wood.

2.6 million years ago The Oldowan toolkit, a set of stone tools built according to simple designs, is developed by *Homo habilis*, a prehuman ancestor. This marks the beginning of the Palaeolithic, or Old Stone Age.

1.5 million years ago Some prehuman ancestors learn to make and control fire and use it for protection, heat and light. The use of fire does not become widespread human technology until around 125,000 years ago.

70,000 BCE Modern human migrants spread from Africa to southern Asia and Australia – and most of the major islands in between. This suggests raft and possibly boat technology has become well advanced.

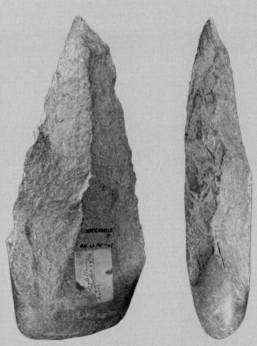

An Acheulean hand axe from Haute-Garonne, France.

EVENTUAL INNOVATIONS

This most primitive mode of stone technology is known as the Oldowan toolkit, for the Olduvai Gorge in Tanzania, where many of the earliest hominid fossils (and their tools) have been unearthed. About 1.7 million years ago it was updated to the Acheulean system, which was used by *Homo erectus*, the first human-like species to spread far out of Africa. Their tools were more refined, with smaller flakes "knapped" off cutting edges to create a sharper and more symmetrical pear-shaped tool.

BOAT TRAVEL

The fact that metal boats weighing thousands of tons can float is still baffling to the uninitiated – although Archimedes did explain it as long ago as 250 BCE. Nevertheless, the first watercraft were obviously made from wood and other biodegradable materials, and therefore our record of their history has been lost. The oldest surviving boat is a 3-m (10-ft) dugout canoe dug up from a Dutch peat bog in 1955. It is around 10,000 years old.

It may be that this craft represents an innovation in boat design from that time – hollowing out a straight log to create a canoe. However, it seems unlikely. Modern humans spread out from Africa at least 70,000 years ago and probably long before, and the speed and pattern of their migration suggests they travelled mostly by water, hugging the coasts of East Africa, Arabia, and South Asia. Modern humans were living on the islands of Indonesia and in Australia long before they spread into the vast hinterlands of Eurasia – and 40,000 years before anyone got to the Americas. Add to this the fact that Homo

Dugout canoe

erectus, a prehuman relative, was living in Asia more than one million years ago and was on Indonesian islands 600,000 years before modern humans had evolved back in Africa.

What form did primitive watercraft take? Natural rafts of floating tree trunks are likely to have been the inspiration for the first craft. As well as the dugouts, which is a traditional boat design the world over, other early designs include coracles, where animal skins are pulled over a light frame, and reed boats made by lashing together long woven bales. This latter design is still the tradition in Andean and Arabian communities.

A BURNING ISSUE

While stone tools are the lasting legacy of the Stone Age, the greater innovation was the taming of fire. How and when this happened is a matter of guesswork. Our ancestors would have been well aware of the characteristics of fire from natural conflagrations started by lightning, and there is evidence of controlled use of fire ranging back 1.5 million years. Friction would have been the main

c. 30,000 BCE Humans start wearing clothes

25,000 The earliest known earthenware artefacts – ornamental figurines – are made in Central Europe.

20,000 Megalith, literally "big stones" are first used in construction in what is now Turkey. This practice spread throughout the world.

13,000 Japanese inventors use pottery to make the first bowls and other vessels for holding water and other items.

10000 Agriculture, where human communities grow their own plant and animal food rather than gather and hunt for it, appears in the Middle East.

9000 Copper is the first element mined by humans, at first in a native pure form and later smelted from ore.

7500 The first cities are built in Mesopotamia. The best preserved is Çatalhöyük in central Turkey.

The creation of fire using the friction generated between two pieces of wood.

method for making fire at will. Sparks can be made by hitting rocks together – we still call the steel sparking fire-lighters "flints" today – and famously "rubbing sticks" together is another kind of method.

By around 200,000 years ago, the presence of ash-filled hearths and fire pits makes it clear that early humans were now ubiquitous users of fire for warmth, light, protection, and cooking, all innovations that we rely on totally today. The next big step was using fire to manufacture artificial materials: around 25,000 years ago people had learned to make stone-like pottery from soft, mouldable clay.

Placing these objects inside a hot fire made the soft clay turn into rock-hard objects. This is not simply a case of drying clay; the heat chemically alters the minerals in it, transforming the pliable substance into a rigid structure. At first, pottery had ornamental or ritualistic uses, but by around 15,000 years ago, people were building functional objects, such as bowls and jars. Pottery of this kind for transporting and storing water and food would have obvious ensuing benefits in cultures making the switch from hunting and gathering to agriculture, plus the act of firing clays also had another unseen result — the advent of metalworking.

The Venus from Dolni Vestonice is at least 27,000 years old.

THE FIRST MACHINE

The hand axe, a fist-sized piece of stone shaped into a pointed blade, was invented more than two million years ago. It is the first machine in history. A machine is a device that modifies a force by changing the direction of its action and magnifying or reducing its effects. There are really only six simple types – the wedge, lever, wheel, screw, ramp, and pulley – and all other machines (a crane, loom, or engine) are just combinations of one or more of them. The hand axe is a wedge. As the idiom explains, this has a thin end and a wider end. A force applied to the wide end is focused into thin, sharp edge, where it becomes strong enough to cut through other materials. The wedge underlies all primitive cutters and weapons, such as spearheads. The knife is its modern equivalent. It is likely that prehuman societies used the leverage of sticks and bones when digging holes and the like, but the other simple machines have only been in use for a comparatively short period – 10,000 years or so!

Prehistoric hand axe

Mystery Megaliths

Stonehenge

The most obvious legacy of the Stone Age is the megalithic monuments, such as Stonehenge in the UK or Carnac in France. Megalithic architecture dates from around 8,500 BCE. Its most basic form is the dolmen, a chamber – generally rather small – created by laying a roof of flat stones on upright rocks.

The word "megalith" literally means "very big stone". At first glance it looks like the stone elements of megalithic structures are simply big rocks that have been placed upright or lain on one another. However, the stones are universally "dressed", which means they have been shaped and smoothed using tools – in many cases stone tools. Megaliths are often connected by joints, such as where protruding tenons on one stone slot into mortice holes on another.

Quite why people, mostly in Western Asia and Europe, started to build in this way is unclear. As well as using stone, the original buildings included wooden structures and earthworks now largely consumed by the landscape. There is no doubt that such buildings were very important, with millions of man hours used in their construction. Stonehenge, the most world's famous megalithic building, is aligned with the motion of the Sun through the year. Perhaps it was some kind of temple or an early astronomical observatory – more likely both!

Compared to heavy but brittle stone and flimsy and short-lived wood, metal is a wonder material. It can be shaped easily, is able to flex a little but maintains its shape. However, it is very rarely seen in the natural world. Of the 90-odd elements (the simple substances of the Universe, more on which later) that occur on Earth only nine of them are native, meaning they appear in a naturally pure solid state, and only three of these are metals: copper, silver and gold.

Pre-Columbian gold artefacts.

GLISTENING METALS

Ironically the easiest to find in ancient times was gold, which would have twinkled in the sandy beds of streams. Pure copper and silver were also located in cracks and seams within rocks where warm, chemically rich waters once trickled leaving behind a deposit of metal.

The first firm evidence we have of gold being used by humans comes from the Varna civilization based in Bulgaria around 6,300 years ago. However, gold has been found in and around human settlements dating back 40,000 years, but whether this is by chance or by design is unclear. It seems likely that gold has a very ancient history indeed. The metal's lasting appeal is based on the way it lasts. Unlike copper, iron and even silver, gold is so unreactive that it is untroubled by corrosion. Gold has always been used to make objects of sacred significance or symbols of high status, and these have not corroded into dust over the generations. As a result gold still is a reserve of wealth that can be relied on to stay unchanged and maintain its value.

A pre-Columbian copper artefact.

6500 BCE Lead smelting begins in Asia Minor. Gold is also widely worked in Egypt and the Middle East, although was probably being used long before this.

6000 The ard, a simple form of plough, is invented to be pulled by oxen or another beast of burden. It is made of wood or deer antler metal.

5000 Copper smelting is developed in the Balkans and spreads quickly to Turkey and Mesopotamia.

4200 Charcoal fuel is used to smelt copper and tin ores and manufacture bronze in Egypt and Sumeria.

3500 Glass is manufactured in Mesopotamia by heating the silicon dioxide in sand.
• The first zoo, or menagerie, is set up in Hierakonpolis, Egypt.

3200 The first wheeled vehicles are invented in Mesopotamia, pulled by tamed wild asses and donkeys.
• Pure iron from meteorites is used as a source of metal in Egypt.

3000 The first maps of constellations made of the brightest stars are formulated by the Sumerians, in what is now Syria and Iraq.

2800 The first ziggurats, large stepped platforms made from mud bricks, are built in Mesopotamia.

2618 The Pyramid of Djoser is the first pyramid built in Egypt.

2550 The Pyramids of Giza in Egypt are built.

2500 The main circle at Stonehenge in England is constructed to mark the solstices, the longest and shortest days of the year.

2296 Chinese astronomers make the first record of a comet.

2137 The first record of a solar eclipse is made in ancient Chinese chronicles.

2000 Babylonians develop a sexagesimal, or base 60, counting system. This way of counting is still reflected in the measurement of time and angles.

1450 Ancient Egyptians divide the daylight into ten hours, and add two more for the periods of twilight at dawn and dusk, thus making a twelve-hour day.

THE WHEEL

Whether it is the most significant invention in history is debatable, and the wheel had much more humble beginnings than many suspect. The first wheel and axle rolled into history around 3,500 years ago. At this time heavy loads were dragged on sleds, sometimes run over rollers, however, it was a while before wheels were added to sleds to make carts and wagons, the first wheeled vehicles. The first wheels were used by potters, spinning around to make it easier to coil clay into bowls.

The first wheeled carts were used in 3500 BCE in Sumeria. Chariots were developed soon after.

THE DAWN OF AGRICULTURE

For most of human history, our species has survived by hunting and gathering, a lifestyle that required frequent movement to follow prey animals and find new sources of plant foods. One of the foods we collected were the grass grains that grew in large meadows. The ears of wild grass shatter when touched. That means the seeds scatter to the ground, well placed to sprout the following spring. However, this was hard work for us humans, forced to pick grains from the dust. Some ears did not shatter quite so easily, however, and the grains could be harvested all at once. The first farmer simply planted the seeds of the non-shattering grasses and grew their own meadows. The following autumn they could harvest the grains with ease. Agriculture had begun. It is thought that this kind of process happened several times independently, first with wheat in Mesopotamia around 10,000 BCE, then rice in east Asia in 8000 BCE, and maize (or corn) in Mexico around 6700 BCE.

A detail of a painting on stucco in the tomb of Sennedjem showing the tomb owner ploughing and his wife sowing flax or wheat seeds.

SMELTING

The total amount of pure gold that has been dug up from Earth's rocks since the beginning of time would make a cube 25 m (80 ft) tall – and fit snugly into the penalty area of a soccer pitch. About a quarter as much is still down there somewhere underground waiting to be dug up. Native copper and silver are more abundant in rocks but we chose to leave it there because, around 8,500 years ago, people living in what is now central Turkey hit upon a technique for extracting metals from the mineral content of rocks.

This method, known as smelting, makes pure metal appear from crystalline solids. It was probably encountered by chance during pottery firing. Potters who added attractive coloured crystals to their pots found the heat from the kiln transformed them into pellets of metal. By loading it with the right crystals – today we'd call them ores – a potter's kiln could become a metalworker's furnace.

A Bronze-Age helmet.

REDUCTION

Smelting is a chemical process known as reduction, which implies that the ore is being "reduced" into its pure base metal. How this reduction might actually work was a focus of the alchemists and early chemists until the late 1700s. Today, we know that ores are generally metal oxides or sulphides, where the metal atoms are bonded to oxygens or sulphur atoms. During smelting, the oxygen or sulphur in the ores is reacting with carbon in the wood fuels and the ore gives up its metal as other constituents form a compound with carbon. The most obvious example would be an oxide ore turning into a pure metal and carbon dioxide.

Modern chemists still use the term reduction. It is a general concept where a substance has its oxygen or a similar element removed during a reaction. The opposite process, where a substance has oxygen or similar added, is called oxidation. In smelting, therefore, the ore is reduced and the carbon fuel is oxidized.

SOFT AND SIMPLE

The first metal to be smelted was lead. This element is less reactive than many metals and so is easily reduced from its ores, requiring lower temperature fires. The most striking lead ore is galena, which is a dark lustrous mineral composed mostly of lead sulphide. Around the year 6500 BCE galena was being smelted in furnaces at Çatalhöyük in what is now Central Turkey. Lead appears not to have been all that useful for the neolithic societies that made it, and it may have been a byproduct of silver working. Pure lead is very soft and no good for making tools or weapons. (The Romans found a use for lead in pipes and waterworks. The Latin for lead is *plumbum*, from which we get the modern word plumber.)

MIX AND MATCH

Partially burning wood so it becomes charcoal, a near-pure carbon fuel, allowed metal workers to boost the heat of their smelters, extracting larger yields of metal. This improvement opened up the possibility of smelting other metals, most notably copper. By 5000 BCE, copper was the main metal being produced around the Eastern Mediterranean. Pure copper is tougher than lead and revolutionized military technology as it was used to make weapons and armour, although stone technologies were still important.

That began to change about 1,000 years later, when pure copper was mixed with arsenic and tin to create a mixture of metals, or alloy, known as bronze. Alloying gives metals greater strength but does not diminish other qualities. Metal atoms are bonded in such as way that they are free to move past each other but not break away, and this underlies a metal's ability to be shaped and bend without breaking. Alloying adds in atoms of different sizes, large tin atoms among the smaller copper in the case of bronze, and this locks the atoms in place creating somewhat greater strength.

Bronze weaponry could be made sharper and tougher than its copper equivalents, and so bronze-wielding cultures were soon on the march, spreading

ROPE STRETCHERS

Agriculture transformed human society in countless ways. One was to create the concept of land ownership. After the annual flood of the River Nile swamped the banks with its fertile muds, officials known as harpedonaptai or "rope stretchers" carefully marked out each family's field. The harpedonaptai used ropes marked with 12 equally spaced knots. Forming these ropes into a triangle with three, four and five knots on each side created a perfect right-angled triangle for marking the square corners of plots. This showed that the Egyptians understood Pythagoras's famous triangle theorem over 2,500 years before the Greek mathematician was born!

An Ancient Egyptian land surveyor.

Bronze Age technology across the world. The Bronze Age lasted around 2,500 years until the crucial supplies of tin began to dwindle. The resulting innovation would not save the technology, but put bronze firmly into the history books.

Bronze-Age weapons from Finland.

The Bronze Age was a time of legends. It was when the events that inform our myths and legends were taking place, such as great floods, tribal migrations, terrible wars, and city-destroying cataclysms; the time of Helen of Troy, Gilgamesh, and the Minotaur.

The great Bronze Age civilizations, such as the Minoans of Crete, Babylon, and ancient Egypt were centred around the Eastern Mediterranean and Persian Gulf. This area became the crossroads of the world's trade routes, a position that has had lasting consequences ever since.

THE MYSTERY OF ATLANTIS

One of the factors behind the Bronze Age Collapse is thought to have been a catastrophic volcanic eruption, which destroyed most of the Greek island of Thera, *also known as Santorini, in 1627 BCE. This island hosted a Bronze Age city called Akrotiri, which was part of the Minoan civilization centred on Crete to the south.*

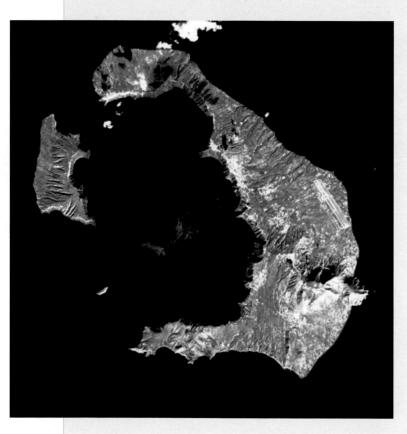

The city was totally buried in volcanic ash. Meanwhile a tsunami from the eruption raced across the Aegean sea to Crete and wrought havoc, perhaps destroying the maritime culture's fleet of ships, leaving them cut off from trade and unable to defend themselves. This story of destruction is the leading contender for the true life events that inspired Plato's story of Atlantis, the civilization that was swamped by the ocean.

The volcanic eruption completely destroyed the middle of the island of Thera.

BRONZE AGE COLLAPSE

However, around 1200 BCE, many of these great civilizations rapidly fall away from history. No one is quite sure why. Perhaps it was climate change, perhaps invaders, especially from the west and north, harried the cities and trading routes, but another factor was certainly the lack of tin, an essential ingredient for making bronze. The tin came mostly from western and northern Europe – a troublesome region – and without fresh tin, armies were forced to recycle whatever they could find. Metalworkers were also running out of trees to supply the raw materials for the charcoal fuel, and so the bronze trade more or less ceased.

The search for new tin sources took 300 years and was ultimately successful, but by that time a new metal technology had taken its place — ironworking. Although iron is considerably more reactive than copper or tin, and therefore requires more energy to be smelted, once pure it can outstrip any copper alloy for strength and toughness. In addition, iron ore is considerably more common than copper and hugely easier to find than tin. So despite the high energy cost to refine it, iron turned out to be a cheaper material than bronze because ore did not need to travel large distances from mine to smelter. The Iron Age had well and truly arrived. The metal, although much refined and alloyed, is still the main one used today for everything from paperclips to skyscrapers.

1300 BCE Ironworking begins in southeastern Europe as Bronze Age civilizations begin to collapse.

1000 Persian towns and cities develop a system of underground channels, or *qanats*, to bring cool, fresh water to houses.

763 Babylonian astronomers record a total solar eclipse.

700 Triremes, warships powered by three decks of rowers, are developed by Greek naval forces in the Eastern Mediterranean.

600 The Cloaca Maxima, the main sewer of Rome, is built forming one of the earliest drainage systems in the world.

585 Thales of Miletus emerges as the first named scientist. He uses astronomical data to predict the date of a solar eclipse. He also studies the properties of magnets and static electricity, and he proposes that the Universe is made from different forms of water.

580 Anaximander, a Greek philosopher, suggests that planet Earth is a cylinder floating in space.

500 The Pythagoreans, a cult-like group of mathematicians, suggest that evidence shows that Earth is a sphere.

• Xenophanes suggests that life on Earth suffers from a string of catastrophes, where life is wiped out by one disaster after another and then regenerates.

Knossos in the Mediterranean island of Crete was the capital of the Bronze Age Minoan civilization.

USING IRON

The iron used by metalworkers actually predates the Bronze Age. Burial goods unearthed from tombs built in the Nile Valley in 3200 BCE used meteoric iron, which as the name suggests is pure iron that arrived from space! There was no need to smelt this material but there was not enough of it to form the basis of a new technology. That began with the first iron smelters. There is evidence of a few successful attempts to purify iron from ore in Mesopotamia in the 3rd millennium BCE, and smelters were certainly operating in the northern Balkans, in what is now Serbia, in 1300 BCE, a few decades before the start of the Bronze Age Collapse. There are also iron workings from 1500 BCE in Niger in west Africa, dating from around the same time that copper and bronze technologies were being introduced to the region.

However, large-scale ironworking required a new kind of smelter, known as a bloomery. This was invented in the Near East; the earliest evidence for one comes from the Jordan Valley. The bloomery relied on bellows, a new invention around 900 BCE, to blast air into the furnace to boost its temperature. The resulting "bloom" was a spongy mix of iron and impurities known as slag. The bloom had to be heated and hammered to drive out the molten impurities, and the result was a very pure form of metal called wrought iron. This was long lasting, easily worked and flexible, which made it useful for some applications but not for others – wrought iron blades and weapons were not very sharp, for example. So a skilled blacksmith would turn the bloom into steel, an alloy or iron and carbon that is still a significant material in modern engineering.

Iron smelting.

STEEL MAKING

While wrought iron had about 0.5 per cent carbon mixed into the iron, steel has up to 2 per cent. The addition of carbon crystals in the lattice of iron atoms allows the metal to flex just a little but also prevents it from cracking when under tension. It takes great skill and a lot of hard work to create carbon steels by hand, and the technology changed slowly. The best technique developed in southern India around 300 BCE, and passed along trade routes to Damascus, in today's Syria. For the next 1,500 years, knights, noblemen, and other professional killers ensured they were armed with swords made from the finest and sharpest Damascus steel.

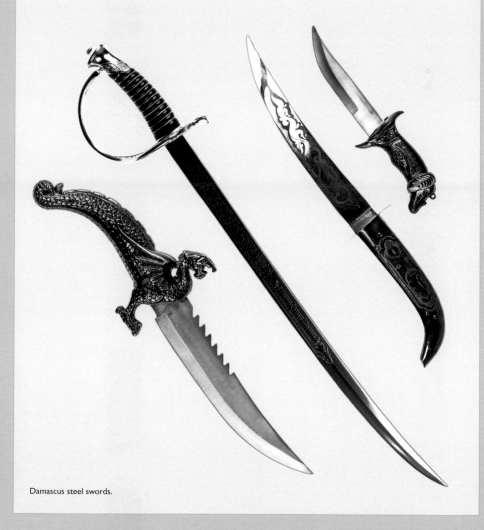

Damascus steel swords.

Ironworking only reached Australia and the Americas during the Age of Exploration that began in the 1400s CE, but the technology had spread to all corners of the Old World – from Ireland to Japan and southern Africa – around 2,000 years before that. Civilizations began making great strides in all regions, most notably in west Africa, China, India, and southern Europe (where the Romans were becoming a force). However, back where it all began around the rim of the Eastern Mediterranean, two cultures dominated: the Persians, who held a mighty empire, and the ancient Greeks, who persisted in a patchwork of city states.

NEW WAYS OF THINKING

Despite all the feats of conquest, new technologies and great art being produced across the globe, the story of science really begins in Greek culture. Quite why Greek culture gave rise to this new mode of investigating the world is open to question. A strong influence is the religion of Classical Greece, which was based on the Olympian pantheon, such as Zeus (the king), Aphrodite, Apollo, Poseidon, and the like. This gang of gods had supernatural powers but were very human in their behaviour, frequently acting out of jealousy, lust and anger. Mortal humans petitioned the gods for favours and protection, but the gods offered few answers to the big questions, such as where did the Universe come from and how does nature work? As a result, mortal thinkers had a go at figuring out those answers themselves, using whatever evidence they could muster. Such an endeavour became known as natural philosophy.

THE FIRST SCIENTIST

Thales is the first person recorded in history for investigating the natural world. Living in Miletus, a Greek city on the western coast of what is now Turkey, he was the natural philosopher's natural philosopher. Thales was the first person to describe the features of electricity. He investigated how rubbing amber made it attract dust

470 BCE The Greek philosopher Anaxagoras proposes that life arrived on Earth from space via meteorites and comets. This idea is now known as panspermia.

460 The classical Greek elements – earth, fire, water and air – are defined and expounded by Empedocles in his poem *On Nature*.

445 Leucippus of Miletus proposes that the Universe is made from indivisible units called atoms.

432 The Parthenon, the main building at the Acropolis, is built in Athens, Greece.

400 Plato suggests atoms are geometric solids with regular faces and edges. There are five of these Platonic Solids, one for each element, and a fifth for "spirit".
• Yakhchals, or conical ice houses, are connected to the qanat system in Persia to make and store ice.

350 Eudoxus, a pupil of Plato, describes the celestial sphere, with Earth at the centre.
• Aristotle suggests that Earth and other heavenly bodies are spheres.

330 Aristotle develops a method of classifying animals by their anatomy.

325 Pytheas, a Greek explorer, suggests that tides are caused by the pull of the Moon.

c. 300 Waterwheels are developed in Greece and later the Roman Empire before becoming the dominant power supply across the ancient and medieval world.

c. 300 Theophrastus, a pupil of Aristotle, founds the field of botany.

270 The Greek astronomer Aristarchus disagrees with Plato's geocentric model and claims that the Sun is at the centre of the Solar System.

250 Chinese researchers note that falling bodies move at a constant velocity.

240 The first record of Halley's Comet is made by Chinese astronomers.
• Archimedes develops hydrostatics, commonly known as the Archimedes Principle, which explains why objects float or sink.

206 South pointers, or primitive compasses, are invented in China.

194 Eratosthenes calculates the size of Earth using the angle of the Sun above different points on the surface.

150 Hipparchus invents the astrolabe, a device for measuring the position of stars in the sky. He divides the night sky into longitude and latitude.

65 An elaborate mechanical gear system called the Antikythera mechanism is lost overboard in a shipwreck near Crete. Its precise function is a mystery but it is thought to be a primitive mechanical computer for predicting the motion of heavenly bodies.

46 The Julian Calendar, named for Julius Caesar, is introduced to correct an inherent error in the Roman calendar. It forms the basis for the Gregorian calendar used today.

Thales of Miletus.

and feathers and it could even release a tiny spark. Thales saw this as evidence of an underlying principle in all material, although, quite understandably, he did not get close to the true description of static electricity (he thought it was linked to water). A better understanding would have to wait until the 18th century CE, although Thales did influence the eventual name for the phenomenon. The Greek word for amber is *elektron* and this was the inspiration for the term "electricity".

31

MATHEMATICAL GROUNDING

According to the Greek historian Herodotus, who lived a century after Thales, the great thinker had predicted a total solar eclipse, another first in history. It is assumed he did this by analyzing the motion of the Sun and Moon, but some authorities question if he would have been able to do this. The event that is now known as the Eclipse of Thales was on May 28, 585 BCE. That day also marked a battle between the Medes and Lydians, two states not far from Thales's hometown. The two armies were in the thick of the fighting when the eclipse cast them into darkness. Both sides saw this as a bad sign and sued for peace.

Thales had yet more lasting success with his mathematical work. There are theories of triangular geometry that bear his name, but he is now firmly in the shadow of Pythagoras. Born 50 years after Thales, Pythagoras is probably the most famous mathematician of all time,

helped no doubt by the fact that we all learn his theorem of right-angle triangles at school. For those who've forgotten: the square of the hypotenuse (longest side) of a right-angle triangle is equal to the sum of the squares of the two other sides.

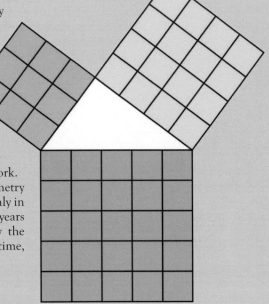

A graphical proof of the Pythagoras Theorem.

$$a^2 + b^2 = c^2$$

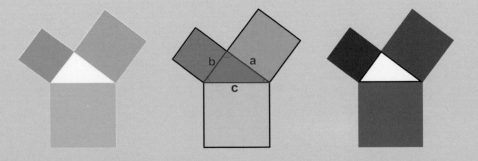

Pythagoras did not make this discovery himself – the rope-stretchers of Egypt had been using it for more than a millennium – but he was probably the first to prove that it was always true. Pythagoras created an entire natural philosophy based on numbers. He believed that the whole of nature, all the processes that produce substances and controlled the way they changed, could be explained by whole numbers alone. Sadly his system came crashing down when one of his followers pointed out a flaw: A simple triangle – the simplest of all – did not conform to plan. Imagine a right-angle triangle with short sides of 1 length unit each. The sum of their squares ($1^2 + 1^2$) is 2. That means the length of the hypotenuse is $\sqrt{2}$. What is that number, a number that results in 2 when multiplied by itself? The answer is 1.41421356... an infinitely long fraction. Such a number was disallowed by Pythagorean philosophy, though it became a major building block in the development of modern mathematics (but that is another story).

Other natural philosophers had a more conventional way of understanding the stuff of nature. All things, they said, were made from mixtures of earth, air, water and fire, four simple materials we now call elements, although the Greeks never did. (This idea was by no means exclusive to Greece either, although Indian and Chinese versions included different materials such as wood and metal.)

Each element has a set of distinctive characteristics. So a soft material was full of air, a hard one was rich in earth, while warms ones contained fire, and so on.

ATOMS

Zeno of Elea, a philosopher from 5th-century BCE *Greece, used paradoxes to expose flaws in theories and ways of thinking. Perhaps his most famous was the Dichotomy, where he sets a runner to cover 1 stadium length. But before the athlete can do that, he must move across half the stadium, and before he can do that he must cross a quarter, then an eighth, sixteenth, and so on, forever. Zeno is wondering how motion over any distance is possible, as it is necessarily made up of an infinite sequence of half distances.*

Zeno's contemporary Leucippus was not having any of this kind of smart-alec thought experiments. He proposed that matter could not be divided into smaller lengths indefinitely. Eventually you would reach a size that was uncuttable – or atomon in Greek. This gave rise to the concept of the atom, a tiny indivisible building block of nature. Leucippus's student Democritus extended the idea, and imagined such things as water atoms being very smooth and slippery and that the human soul was made of ultra-small and lightweight fire atoms. So the idea of an atom was born, but only in a purely theoretical sense. The first evidence of atoms actually existing did not arise until the early 1800s CE.

LAYERS WITHIN LAYERS

Aristotle, perhaps the most famous Greek philosopher, worked in the 4th century BCE, and built on the ideas of his forebears to create a view of the Universe. This was a place of harmony and order, beginning with the Earth, which sits at its centre. Aristotle proposed that nature was in a state of change because the four elements were seeking to separate from one another. Rain and lightning was water and fire escaping from air, while burning was the air (smoke) and earth (ash) in wood separating out. Every element found its own level, beginning with heavy earth which formed the lowest layer. Next came water, then air and finally fire, which formed a ring around Earth just this side of the Moon.

Beyond the fire layer, the Universe was filled with ether, which Aristotle called the fifth element, or quintessence. Unlike the four earthly elements, ether was a heavenly substance, perfect and unchanging, as the Sun and the planets (only five were known at this time) moved in circles around Earth. The outer edge of the Universe was a crystal sphere studded with stars.

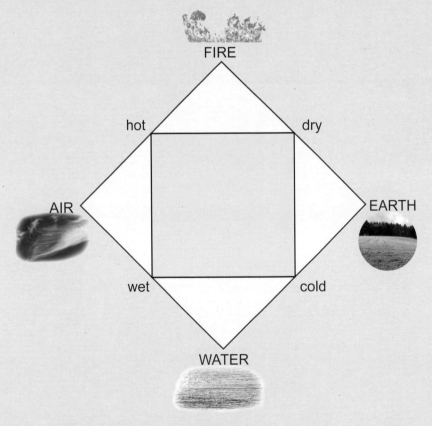

A chart of the classical elements and their supposed properties.

Aristotle proposed that his system must have been set in motion by a prime mover, or "that which moves without being moved," and in Aristotle's view that must be "an immortal, unchanging being, ultimately responsible for all wholeness and orderliness in the sensible world." As a result of this, Aristotle's Universe became a part of the dogma of the Catholic Church. It was heretical to question its veracity, and that would prove a problem when it was found to be mostly wrong.

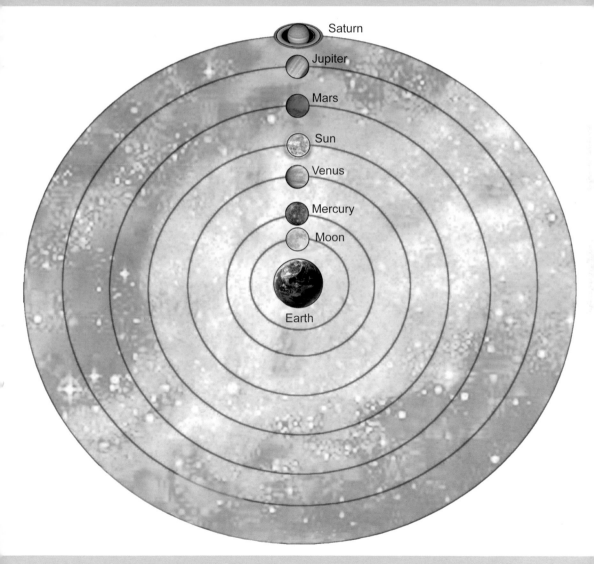

Saturn

Jupiter

Mars

Sun

Venus

Mercury

Moon

Earth

Aristotle's geocentric universe.

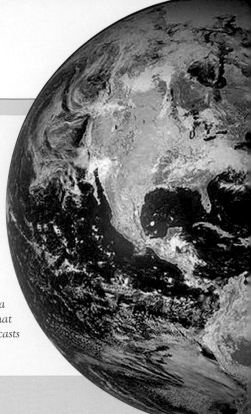

SPHERICAL EARTH

Aristotle may have been wide of the mark in many ways, but his logic was impeccable when explaining the evidence for Earth being a sphere. The first clue was that seafarers always saw the top of a ship's mast before seeing the hull. They suggest that from the point of view of the observers, the ship is rising up toward them. Second, the same seafarers reported seeing different stars when they sailed south, with northern constellations becoming hidden behind the curve of the planet. Third and finally, during a lunar eclipse Earth casts a shadow on the Moon. That shadow is always round. The only shape that always casts round shadows is a sphere.

MEASURING THE EARTH

About a century after Aristotle, Eratosthenes, a Greek mathematician living in North Africa, found a way to measure the size of the Earth. He knew that the Sun was directly overhead the city of Syene (now Aswan, Egypt) on midsummer's day – so much so that it shone right to the bottom of a well without casting a shadow. On the same day Eratosthenes measured the angle of the Sun in the sky from Alexandria in Egpyt, thus creating two lines from the two cities to the centre of the Earth. He knew the angle between them, and calculated that it was a 50th of a full circle. Therefore the distance from Alexandria to Syene was a 50th of the distance all the way around Earth. His final answer was that Earth's circumference was 252,000 stadia, or a distance of 44,100 km (27,400 miles), an error of just 10 per cent.

Aswan well.

ARCHIMEDES

Famous for saying, "give me a lever long enough and a place to stand, and I'll move the Earth", Archimedes made many advances in engineering, using levers and pulleys to move entire ships and reportedly setting fire to enemy vessels with parabolic dishes that focused the Sun's light into a searing beam. In mathematics he was the first to approximate the number pi (the ratio of a circle's radius to its circumference) to 3.14, the figure that has been used in classrooms over the ensuing 2,250 years. However it is what Archimedes did in the bath that bears his name. While noticing how his body displaced water as he sat into a brimming tub, Archimedes realized in a flash what made objects float or sink: if the object weighs more than the volume of water it displaces, then it will sink. If it weighs less then it floats. This is the Archimedes Principle, and it explains why vast metal ships can be made to stay afloat.

Elephantine Island in the Nile near Aswan, where the well of Eratosthenes is located.

Alchemists, part scientist, part wizard, bridged the gap between the classical era of natural philosophy and the Scientific Revolution of the 17th century. Ancient Greek natural philosophers spent their days contemplating nature in the shade of their favourite trees, then talking about what they had thought up. (It was generally their students' job to write it all down, and confusion often reigns over whose idea it was.) Alchemists were much more hands-on researchers, but historians still have a big problem figuring out who did what and how they did it, albeit for very different reasons.

DARK ROOTS

The word "alchemy", which eventually gave us the term "chemistry", has a complex etymology. Primarily it is derived from the Arabic word *al-kimiya*, but this has a more ancient and opaque roots. The Arabic probably comes from a Greek term, *khemia*, meaning "the fusion of metals". As we'll see, alchemists paid particular attention to the behaviour of metals, especially precious ones. However, *khemia* is said to refer to the ancient Egyptian word for black earth, referring to the fertile soils of the Nile Valley. So all in all, we could perhaps settle on alchemy being a mysterious discipline that emerged in Greco-Roman Egypt. There the city of Alexandria (named for Alexander the Great, Aristotle's most famous pupil) had become the world's centre of learning. At the city's great library, knowledge from across the ancient world, China and India especially, mixed with Western thought

An alchemist experimenting in a laboratory.

The great library of Alexandria.

c. 50 CE The first steam-powered engine that converts the expansion of hot gas into motion is invented by Hero of Alexandria.

60 The Pont du Gard, an aqueduct over a river in the South of France, becomes the tallest Roman structure.

75 Hero of Alexandria describes the six simple types of machine: lever, wheel, pulley, ramp, wedge, and screw.

77 Pliny the Elder, a Roman writer, publishes *A Natural History*, one of the first biology books.

105 Paper is invented China from pulp made from hemp.

126 The Pantheon is built in Rome. The concrete dome is 43.3 m (142 ft) wide and it remains the world's largest dome made from unreinforced concrete.

150 Galen, a Greek doctor working in Rome, proposes that living things survive due to an invisible vital spirit that is drawn from the air.

300 Alchemy, a fusion of magic and science that develops first in Egypt, spreads across the world and becomes the dominant form of investigation of nature.

550 John Philoponus of Alexandria proposes that motion is the product of a continuous impetus acting on bodies.

in the centuries before the start of the common era.

The alchemist has informed much of the modern view of a wizard: a hermit figure who worked in secret, muttering spells as he mixed a multitude of strange concoctions. A person with unknown powers, who was not to be trusted. This is fair enough. Alchemists were not in the business of pushing back the boundaries of knowledge; they were in it for the money and power.

THREE GOALS

To alchemists, what we identify today as magic was part of all natural processes. As a result, their goals were somewhat magical. Alchemists worked to find three things: the panacea, the elixir, and the philosopher's stone. The panacea was a substance that could cure all illness. The elixir was a liquid that gave the drinker immortality. Most important of all was the philosopher's stone, which could turn inexpensive base metals (such as lead and iron) into gold. Discovering any of these items would have led to untold riches, and so alchemists kept their work deliberately obscure lest a rival steal their findings. As a result, charlatans found it easy to swindle the credulous and desperate with promises of great wealth and health.

The search for the panacea and the elixir continues to this day, but it is now technically possible to manufacture gold from other substances – though the costs are so astronomical that no one ever does. The modern chemist's understanding of gold is actually very different to that of an alchemist, but that understanding owes itself to work carried out by these olden-days researchers.

CHEMICAL LEGACY

Alchemists developed many of the techniques still used in today's laboratories, such as distillation, filtering, and precipitation. Much of the language they used remains in use. Hard solids were called stones, while the crumblier ones were earths and salts. Today the term "rare earths" refers to a set of elements, such as neodymium and erbium, which are hard to find but technologically crucial. Wet liquids were referred to waters (they probably were mostly water). Thicker ones were oils, while the most volatile liquids were called spirits, because they were thought to contain some concentrated essence of a substance. Alchemy also produced some tangible discoveries, such as alcohols, dyes and several seemingly elemental substances, such as sulphur and metals that were not on the original short-list of elements. Much of this work occurred while Europe was in its so-called Dark Ages. Elsewhere in the world, however, science was developing in leaps and bounds.

Ancient alchemical apparatus.

PAPER

The Chinese pride themselves on being responsible for four great inventions: gunpowder, the compass, paper, and printing. Paper was the first invention we have evidence for, dating from the start of the 2nd century CE. Chinese paper was made from plant fibres, such as hemp, left over from textile manufacture. These fibres were soaked and then pummelled into a pulp that could be dried out into flat sheets. The ancient Egyptians had been using something similar, papyrus, for much longer. This was made from dried strips cut from reed stems. Nevertheless, paper technology could be expanded to all kinds of plant fibres, including wood pulp which is used today.

Ancient Chinese writing on paper.

Hero of Alexandria

Steam power was invented long before the Industrial Revolution, but its inventor, Hero of Alexandria, only toyed with the idea of using it to drive machines. Instead, his device from the 1st century CE became a curiosity for impressing his friends. Hero called his invention the aeolipile, meaning "wind ball". It was a metal sphere with two pipes emerging on opposite sides. Steam from a boiler filled the sphere, and high-pressure vapour emerged from the pipes, making the ball spin around. It would be another 1,600 years before this rotating system would really put civilization into a spin, however.

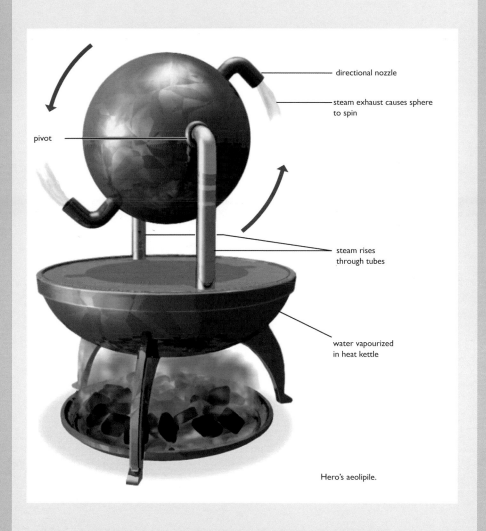

directional nozzle

steam exhaust causes sphere to spin

pivot

steam rises through tubes

water vapourized in heat kettle

Hero's aeolipile.

The fall of the Roman Empire, in western Europe at least, is frequently characterized as the start of the Dark Ages. This period is proposed to be one of cultural reverses, where the hard-won civilization of the Romans was eroded by the brutish invaders that took control of their territory. This is largely nonsense. The Goths, Vandals and Huns, whose very names are still epithets for unsophistication and thuggery, were just as cultured as the Romans (well, perhaps not the Huns). However, the "darkness" of the age is due to the fragmentation of historical records, so we do not know what was going on. Not so to the east, where the burgeoning Islamic Empire was centred on its capital, Baghdad. This city's House of Wisdom had become the prototype university, and so Baghdad took over from Alexandria (its library largely lost to fire) as the centre of world knowledge. What followed was the Golden Age of Islam, and many of the breakthroughs in science were discovered by scholars hailing from anywhere between Pakistan in the east to the Atlas Mountains in the west.

Among many examples are two Persians born in what is now Uzbekistan: al-Khwarizmi who invented algebra, and al-Biruni who calculated the radius of planet Earth and reasoned that the world was so large that there was probably other vast continents still to be discovered. Meanwhile, Ibn Rushd (also known as Averroes) from Spain was considering the concepts of acceleration, force, and inertia 500 years before Isaac Newton put them all together in his Laws of Motion.

House of Wisdom, Baghdad.

MERCURY–SULPHUR THEORY

One of the earliest contributors was the Persian alchemist Jabir ibn Hayyan. By his day in the 8th century, alchemists were will acquainted with the idea that nature contained more than just four primordial substances. These included carbon and a range of metals, but Jabir (also known by the Latinized form Geber) was most interested in mercury and sulphur.

In alchemical terms, Jabir saw both as spirits, in that they encompassed an essence of nature. In medieval days, sulphur was better known as brimstone, which means "stone that burns." This yellow solid forms in crusts around the edge of hot pools and steaming volcanic vents. When set alight, the solid melts into an oozing blood-red liquid which burns with a pale blue flame. Still quite a sight when seen in the laboratory, when it occurs "in the

750 Jabir ibn Hayyan, known in Europe as Geber – a Persian alchemist – classifies substances as metals, spirits, and powders, or earths.

800 Zinc identified by the Indian alchemist Rasaratna Samuccaya.

c. 850 Gunpowder, the world's earliest explosive, is invented in China, reportedly by accident as alchemists searched for the elixir of life.

990 Al-Biruni, an Islamic geographer, calculates the circumference of Earth using measurements taken from a mountaintop in India.

wild" it has a lasting impact on those that see it. The ancients saw brimstone as the fuel that fed the fires of the underworld.

Mercury was originally named by the Greeks as "water-silver" (later to be *hydragyrum* in Latin), to reflect its liquid and metallic qualities. It could be made by roasting a dark red mineral called cinnabar. A later name for the metal was quicksilver, which echoed how the liquid flowed in a fast and erratic fashion. Today's name is taken from the messenger of the Roman gods, who was also fast and erratic (so too is the planet Mercury, which dashes across the sky.)

Burning sulphur.

Abu Musa Jabir ibn Hayyan, Persian alchemist and father of early chemistry.

TALKING GIBBERISH

Jabir proposed that mercury gave metals their shine and malleable qualities. His theory also held that each type of metal – gold, lead, etc. – contained different amounts of sulphur as well. The precise mixture gave them their varying qualities, such as gold's inert shine. Nevertheless, Jabir admits that he is not being literal here. Mixing mercury and sulphur does not make gold or any other metal. Instead it will create an artificial form of cinnabar, among other things. So Jabir is referring to a metaphorical or philosophic essence that he sees as being embodied within mercury and sulphur. It took several centuries more to unpick these ideas, but essentially he was attempting to describe concepts such as energy and reactivity. Nevertheless, in keeping with his alchemical background, Jabir obscured his theories in a baffling code of symbols that no one else could understand. Jabir's bad writing earned him a place in the English language: the word "gibberish" is derived from his name.

Jabir ibn Hayyan.

1010 The Arab scientist ibn Al-Haytham proves that vision is the result of light arriving from objects, not the reflection of rays coming from the eye. His work on light forms the foundation of optics.

1025 Avicenna, a Persian doctor also known as Ibn Sina, publishes *The Canon of Medicine*, which replaces the work of Galen as the leading medical text book until the Renaissance.

1040 The magnetic compass is reported to be used in navigation for the first time in China.

1054 Chinese astronomers record a new, very bright star. This is the first record of a supernova, an exploding star, being observed by humans. The remains of the explosion forms what is now called the Crab Nebula.

1121 Al-Khazini of Merv, now in Turkmenistan, suggests that gravity is a force that pulls objects to the centre of Earth.

1150 Averroes defines force by the rate of change in the motion of an object. His ideas on resistance is an early description of inertia.

1250 Albertus Magnus, an Italian alchemist, isolates the element arsenic and helps to develop an empirical approach to alchemy.

1256 *De Vegetabilibus*, a catalogue of herbs and other useful plants, is published by Albertus Magnus.

1267 Roger Bacon investigates the way lenses can magnify objects.

1304 Thierry of Freiburg explains how rainbows are caused by the refraction of light in raindrops.

1320 The canon of parsimony, which says the simplest explanations are generally the right ones, becomes known as Occam's Razor after being reiterated by English philosopher William Occam.

1350 Jean Buridan preempts Newton by saying that objects slow down not because their motion is dissipating but because an opposing impetus is pushing against them.

MAKING MORE SENSE

The following century, Abu Bakr Muhammad ibn Zakariyya al-Razi, better known as Rhazes or simply al-Razi, brought some much-needed clarity to alchemy (and a range of other fields). As a result, chemistry still uses Arabic words. "Alkali", denoting a substance that reacts strongly with an acid, was coined from the Arabic *al-qaly*, meaning calcinated (or roasted) ashes. In addition, "alcohol" comes from the term *al kohl*, the dark eye make-up worn since ancient Egyptian times. This was a finely powdered mineral, which al-Razi saw as being, in a basic form, the spirit of the substance. And so too were the volatile liquids that he distilled from fermenting mixtures. Islamic society does not use alcohol as an intoxicant. Instead alcohol was a powerful solvent for plant extracts, creating a perfume industry that fed markets in the East and West alike for centuries.

The modern word "alcohol" is derived from the Arabic term *al kohl*, the dark eye make-up.

45

SEEING IS BELIEVING

At the dawn of the second millennium, in the year 1011, Ibn al-Haytham was summoned to Cairo by the powerful Fatamid caliph. Al-Haytham (or Alhazen) had foolishly boasted that he could orchestrate the damming of the Nile, a feat beyond even a man of his abilities. Facing jail (or worse) for messing the caliph around, al-Haytham feigned mental illness and so was let off with house arrest – for the next ten years! While in detention, al-Haytham became interested in the behaviour of light and in particular the camera obscura, literally a dark room where light entered through a single point in one wall creating an inverted image of the outside on the opposite wall. Al-Haytham assumed that light beams always moved in straight lines, and so was able to use geometry to explain how they behaved. This made him the founding figure of optics, and his most significant discovery was that vision was the result of light from the Sun or another source, a candle let's say, entering the eye. Before this, most scholars believed that the eye sent out invisible beams to scan the surroundings like some corporeal radar system.

The typical blue of Middle-Eastern tiles comes from dyes developed by alchemists.

COMPASS

The first compasses were naturally magnetic iron-rich minerals called lodestones. The Chinese were the first to notice that when left to move freely, these stones always pointed in the same direction. The Chinese called their compasses south pointers – that direction was obviously more appealing to them. The first south pointers were being used as long ago as the 3rd century BCE, though they were not used for navigation but for geomancy, or aligning buildings in auspicious directions. The first evidence of a navigation compass was from 1040 CE. It was a lodestone flake carved into the shape of a fish so it could float on water and swing to the south. Once compasses reached Europe in the 14th century, they were made to point north.

Chinese geomancy compass.

Navigation compass.

GUNPOWDER

Also called black powder, gunpowder is simply a mixture of powdered charcoal (carbon), sulphur, and saltpetre. It was invented in the 9th century CE by Chinese alchemists reportedly in search of the elixir of life. They found the opposite because they discovered the world's first explosive, a substance that burns all by itself without the necessity of air or another external substance. Saltpetre is a nitrate, and its molecules split open releasing oxygen, which drives the combustion of the other two elements. Once the powder is burning, it will not go out until that fuel is extinguished, and it will burn so fast that it creates a shockwave of high pressure and heat. That explosive power was initially used for fun in firework rockets, but by the 13th century, black powder was a military technology across Europe and Asia.

Working in the mid 20th century, the American philosopher Thomas Kuhn coined the term "paradigm shift." He used it to define the conclusion of a crisis in scientific thinking, or a period where the sheer weight of unknown mysteries and contradictions weakened the accepted scientific thinking until it broke. What emerged was a new way of understanding nature that relied on a new set of assumptions – the paradigm had shifted.

SUN OR EARTH?

The greatest paradigm shift of all moved the Earth – not literally but figuratively – in the minds of every human who has lived ever since. The man responsible was Nicolaus Copernicus, a Polish clergyman with an interest in mathematics and astronomy. In 1543 he presented evidence that Earth orbited the Sun as one of the planets and not, as was the accepted view of the time, that Earth was the centre point of the Universe circled by all other observable objects, including the Sun.

Copernicus was not the first to suggest this heliocentric, or Sun-centred, model

Nicolaus Copernicus.

1404 Henry IV outlaws alchemy because he is afraid that alchemists will discover the philosopher's stone and flood the country with gold, thus reducing his wealth and weakening his hold on power.

1440 Nicolas Cusanus suggests that Earth is in motion through space, rejecting the classical idea that the planet is in static perfection.

1450 Bismuth and antimony are described as distinct elements by Basilius Valentinus, although the metals have been used in alloys with lead and tin for centuries.

• Ibn al-Nafis makes the first accurate record of the circulatory system of the human body.

1499 Leonardo da Vinci makes anatomical drawings including of the human brain – a heretical act at the time.

1500 Venetian shipwrights start producing galleons, which are the largest vessels on the ocean and used for carrying troops and cargo.

1508 Michaelangelo begins to paint the ceiling of the Sistine Chapel in the Vatican, and draws God in the shape of a human brain.

1528 Grisogone of Zara proposes that tides are caused by a magnetic pull from the Moon.

1543 Copernicus publishes details of his heliocentric (Sun-centred) Universe which explains that Earth and the planets orbit the Sun. This overturns centuries of belief in geocentrism, or an Earth-centred model of the Universe.

1550 German metallurgist Georg Pawer, but better known as Agricola, writes *De Re Metallica* (*The Nature of Metals*), a compendium on mining and refining metals.

1551 Conrad Gessner founds the science of zoology, the study of animals.

1555 Pierre Belon finds that the anatomy of many mammals, especially the skeletons of mammals, are made up of the same bone units but are modified for different purposes.

1572 Tycho Brahe observes a nova, the scientific term for a new star seen in the sky where once there was nothing to see. This is one of just a handful for supernovae – less than ten – that have been visible from Earth during human history.

1580 Jan Baptist van Helmont, an alchemist from Flanders, proposes that plants grow by converting water into body tissue.

ANATOMICAL HOMOLOGIES

French naturalist Pierre Belon.

After many centuries of neglect, the life sciences saw a revival in the 16th century. A leading figure was the French naturalist Pierre Belon. In 1555, he published a book on the natural history of birds, but that name belied its true significance. Belon was an avid anatomist and his detailed drawings of bird skeletons of all shapes and sizes revealed something obvious. As well as sharing similarities with other animal species, bird skeletons also resembled the human skeleton. They all had largely the same number of bones arranged in the same order to create the spine, torso, limbs, skull, and jaw. Anatomical differences – wings instead of arms, etc. – were due to variations in the shape and size of the bones. Belon was the first to recognize the homologous nature of anatomies, a discovery that went on to inform theories about how life forms are related, how to classify them, and how they came to be in the first place.

of the Solar System. As far back as about 270 BCE, Aristarchus of Samos proposed the same thing. His book *On the Sizes and Distances of the Sun and Moon* detailed the relative positions of the Moon, Earth and Sun. These results suggested to Aristarchus that Moon and Earth were considerably closer to each other than to the Sun. That also meant that the Sun was much bigger than Earth (although his calculations were very wide of the mark), and Aristarchus viewed the size of the Sun as a clear indication that it formed the centre point of the system. However, without clear evidence, Aristarchus's idea was not able to compete with the Aristotelean view that put Earth at the centre of all things.

Illustration of the Copernican Heliocentric system.

PART OF DOGMA

By Copernicus's day in the 1500s that view still held, if anything strengthened by being an essential component of the received wisdom that formed the world view of the Catholic Church, all-powerful in Europe. Aristotle's universe had been updated and refined by Claudius Ptolemy, a Greek-speaking Roman from Alexandria, Egypt. In about 140 CE, Ptolemy published the most in-depth star map of the age. One shies from describing the work, now known as the *Almagest* (derived from the title given to it by later Arab scholars – *al-majisti*, or "the greatest") as the most accurate to date, because it was based on the great misconception that objects moved around Earth.

Claudius Ptolemy.

50

Hipparchus observing the stars.

TYCHO BRAHE

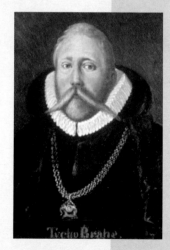

One of the Aristotelean precepts was that the heavens were unchanging and perfect. In 1572, 29 years after Copernicus began rearranging the heavens, his successor as greatest living astronomer noticed something new in the sky. The stargazer was Tycho Brahe, a boorish, brilliant and enormously wealthy Dane, who observed the skies from an underground castle sunk into a Baltic Sea island. Tycho recorded a star as bright as the planet Venus, which appeared as if from nowhere in the constellation of Cassiopeia. It took 450 years to figure out that the star was a supernova, or an explosion produced as a giant star degenerates into a neutron star or black hole.

Tycho rejected Copernicus's theory because he said that if Earth was moving in space, the positions of the stars should shift slightly relative to one another – a stellar parallax. (They do but it is so tiny that it was another 250 years before it could be detected.) Tycho came up with his own model of the Universe, where the Sun and Moon orbited Earth and the other planets orbited the Sun.

Ptolemy had in fact got most of his data from Hipparchus, an astronomer of great skill who worked on the island of Rhodes in the second century BCE. Hipparchus is responsible for many of the innovations that help today's astronomers make sense of the heavens. He used a system similar to longitude and latitude to pinpoint locations in the sky, and classified each star he could see with a score for brightness, now known as magnitude.

Another of Hipparchus's clever contributions was a method of solving the observational mismatch between the way the planets moved and how that tallied with their supposed journeys around Earth. Planets frequently appeared to slow down, speed up, and even change direction when viewed from Earth. To explain this away, Hipparchus built a complex system of wheels within wheels, where planets orbited around off-centre points (eccentrics) and moved in a corkscrew path as they orbited a point that was itself in orbit around Earth, a system called an epicycle.

ADDING COMPLEXITY

Ptolemy checked and updated Hipparchus's map and had to amend the tortuous system of epicycles and eccentrics to maintain an accurate description of planetary motion. The result was so complex and barely workable that by the 16th century, few people understood it well enough to question it. Not so Copernicus.

Copernicus's uncle and patron was a "prince bishop" in northern Poland, which means he was not only the spiritual leader but also the regional ruler. As a result the young Nicolaus wanted for nothing and was educated in the best institutions. However, what he was taught was somewhat limited. The university curriculum was based on Scholasticism, which focused on a narrow set of ideas in keeping with religious dogma. The nearest students got to studying science was the work of Aristotle, and it was frowned upon to question this view of nature. For many centuries, scholars had been trying to find ways of explaining what they observed in nature with the predefined system – and by the end of the 15th century, this way of investigating the world was in crisis.

A NEW VIEW

Multiple advances in mathematics coupled with more precise observational equipment were revealing insurmountable errors in the Ptolemaic order. Copernicus would have heard of such things and confirmed them with his own observations. However, he had a problem. Uncle Lucas, who paid for everything and relied on his nephew to uphold his good family name, had enrolled Nicolaus as an apprentice with Papal Curia, the Pope's civil service. The plan was for Nicolaus to become a priest and rise up the ranks. Therefore, when Copernicus started to consider explaining that Aristotle, Ptolemy and the Pope were totally wrong about the Universe (and he had the observational data to back it up) he faced the wrath of family, the state and God all at once. So he kept it quiet, only sharing his findings with a trusted few. And so it remained for the next 40 years or so as he worked for his uncle and various clergymen

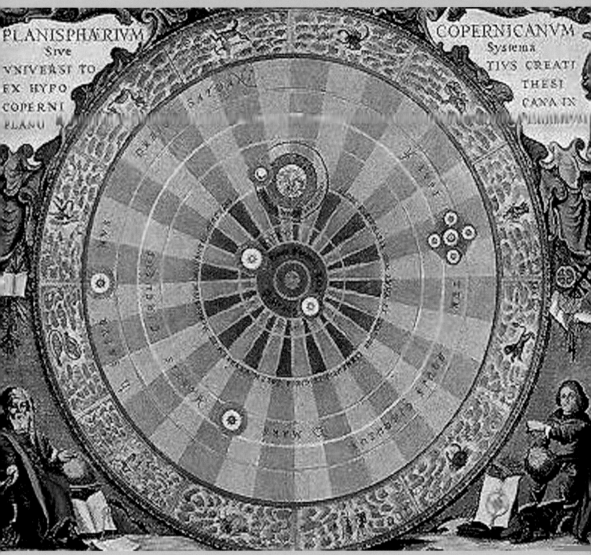

Copernicus placed the Sun at the centre of the Solar System.

in Poland. Only when he lay on his deathbed in 1543 did Copernicus agree to publish his long-held views in a book called *De revolutionibus orbium coelestium* (*On the Revolutions of the Celestial Spheres*). He was dead within a week, and so was beyond the reach of the law for his heresy.

The Church argued that the physical evidence of the rising Sun and changing Moon were unchanged by Copernicus's ideas and no one had even seen anything that was not moving around Earth. However, in a few short decades, an Italian scientist called Galileo would change all that.

I f modern science starts anywhere, it starts with Galileo Galilei. His work was so significant that we don't even bother with his second name anymore. He was the first to explain his discoveries in terms of mathematics, which not only ensured his discoveries were sound but offered them up as component parts in future discoveries.

Galileo Galilei.

PENDULUM LAW

G alileo's long list of discoveries began while he was still a university student in Pisa in the 1580s. While attending mass in the city's cathedral, Galileo noticed that when a verger lit the candles on a lamp suspended above the nave, he would set it swinging. Using his pulse as a rudimentary stopwatch, Galileo found that each swing

Galileo's lamp in Pisa Cathedral.

Galileo shows off his telescope in Venice.

1581 Galileo Galilei reports measuring the period (swing time) of a swinging lamp in church and finds that it is the same no matter how widely it is set swinging. This leads to a mathematical pendulum law.

1582 A new calendar commissioned by Pope Gregory XIII is introduced in much of Europe. The Gregorian calendar adds a day every four years to correct the drift of dates from the earlier Julian calendar.

1593 Galileo is credited with building the first thermoscope, or air thermometer.

1600 English physician William Gilbert shows that Earth has a magnetic field, and this is why compasses always point north.

from side to side took the same amount of time, no matter how wide it was. This formed the basis of the Pendulum Law as presented by Galileo in 1602 – after a lot more testing. This law states that the period (swing time) of a pendulum, such as the cathedral lamp, was defined only by its length. A heavy weight at the end swung at the same speed as a lighter one. A 99-cm (39-in) pendulum has a period of 1 second, which opens up the possibility of using pendulums as a means of measuring time. This is exactly what the Dutch scientist Christiaan Huygens did with the invention of the pendulum clock in 1656.

A MAGNETIC PLANET

In 1600, William Gilbert, an English scientist (and personal physician to Queen Elizabeth I) solved a long-running conundrum: why do magnets always point north? His conclusion was that Earth itself is a magnet. He proved this by carving a model Earth, or terrella, from a lodestone. When he placed a compass on the surface of this little magnetic sphere, it behaved in exactly the same way as on the surface of Earth, showing the planet was effectively a very much larger magnetic sphere. Where the magnetism comes from is still a bit of a mystery even today, but is thought to be something to do with the solid iron inner core spinning within the liquid metal outer core.

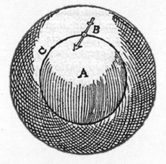

William Gilbert's Terrella.

SEEING FURTHER

After university, Galileo settled down as a teacher at the University of Padua. In 1608, Hans Lippershey, a lens-maker in faraway Middleburg on the coast of the Netherlands, invented the telescope. One version of the story was that Lippershey's children were playing with a pair of spectacle lenses, and found that when they looked through both together they could make the church steeple in the distance appear much larger. Whatever the truth of the matter, Lippershey's telescope rapidly became a sensation among the scientists of Europe. The following year Galileo built his own telescope, grinding the lenses by hand to create a 3x magnification, and presented it to the rulers of Venice. The telescope would be a valuable tool in this mercantile city. Traders would be able to spot approaching ships far from port and use that time advantage to set prices and do deals before goods reached land. The Venetian government offered Galileo a retainer in return for use of the device.

STARRY MESSENGER

Meanwhile, Galileo built himself a much bigger and better telescope with a 30x magnification and pointed it at the stars. What he saw revolutionized astronomy once presented in a short pamphlet called *Siderus Nuncius* (*The Starry Messenger*). The telescope showed that the Moon was a world in its own right. He saw that the terminator, the line between light and dark, was not a straight but made jagged as the Sun was shining on huge lunar landscapes, such as mountains and craters. His telescope also showed that the faint cloudlike nebulae, which looked like pale smudges to the keenest naked eyes, were in fact packed with stars. Galileo could also see more details in the Milky Way, too, but his telescope could not resolve much further that pale, milky streak of light that crosses the sky. Nevertheless, his deduction was that the Milky Way was also filled with so many stars that their light merged. He was not wrong. The Milky Way is our view toward the central bulge of our galaxy, which contains about 200 billion stars in total.

Features of the Moon as shown in Galileo's book *Sidereus Nuncius* or *The Starry Messenger*.

PLANETS ON PARADE

However, most important were Galileo's observations of the planets. Viewed up close, Venus appeared in phases like the Moon. The planet was being illuminated by light from the Sun and appeared to change shape when seen from Earth, clear evidence of the Copernican model of the Universe. Perhaps more significantly, Galileo found four tiny "stars" moved around Jupiter. He named them Medicean Stars – and soon he was employed by the powerful Tuscan Medici family – they were the first clear evidence that not everything in the Universe moved around Earth. We now know them as Jupiter's moons, Io, Europe, Ganymede and Callisto.

The Starry Messenger put Galileo in conflict with the Catholic Church. He

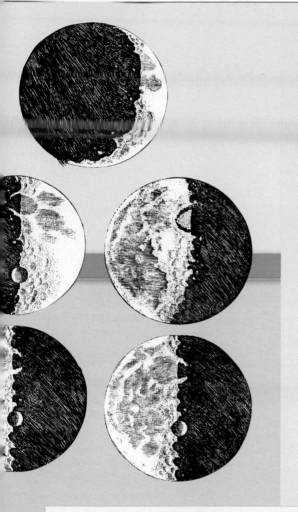

1600 After many centuries of development, the Great Wall of China reaches its largest size, spreading from the Yellow Sea to the Gobi Desert.

1608 The telescope is invented in Middleburgh in the Netherlands. This town is a world centre of lens-making for eyeglasses. The actual inventor is not clear but is generally credited to Hans Lippershey.

1609 Johannes Kepler publishes his *Laws of Planetary Motion* derived from Tycho Brahe's observational data. This work shows that orbiting bodies move in ellipses, not circles.

1610 Galileo uses a large, home-made telescope to offer the best view of the heavens seen to date. He publishes his observations of the Sun, Moon, and planets in a short book called *The Starry Messenger*.

1620 As part of his mission to reinvigorate investigation into the natural world, Francis Bacon publishes *Novum Organum*, with sets out an early version of the scientific method.

1620 Cornelis Drebbel, a Dutch engineer, inventor and innkeeper, builds the world's first submarine. The oar-powered vessel makes a maiden voyage in the River Thames, London.

1621 Dutch astronomer Willebrord Snell sets out a way of calculating the degree to which light bends as it refracts through different media. This is now known as Snell's Law.

1628 The English doctor William Harvey explains the double circulation system of the mammalian (including human) heart.

OVALS NOT CIRCLES

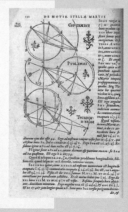

In 1601 Rudolph, the Holy Roman Emperor, appointed the German mathematician Johannes Kepler to turn Tycho Brahe's observations into a star catalogue of such quality that it would be worthy of the title The Rudolphine Tables. Kepler used the planetary data to reject Tycho's cosmological model (where the Sun orbited Earth and the planets orbited the Sun) and confirm the view of Copernicus. In 1609 Kepler was also able to explain the motion of all the planets using one set of mathematical laws. Firstly these laws said that orbits were never circular but ellipses, and secondly, that planets always sweep out a constant area of their orbital disks. For example, each month, Earth sweeps out a 12th of the area of its orbit, although it may move faster or slower each month depending on its position in the elliptical path.

was warned to keep quiet but eventually published his book in 1632. The book was banned (but only until 1835), and Galileo was sentenced to house arrest. In 1638, in failing health (he was dead barely three years later) he smuggled out the manuscript of *Two New Sciences*, a round-up of his life's work, for publication in the Netherlands, where the writ of Rome carried less weight.

LAW OF FALL

Two New Sciences included Galileo's Law of Falling Bodies. His mathematical discovery here was that the distance an object falls is proportional to the square of the time it travelled. Galileo is meant to have tested his idea by dropping a large cannonball and a smaller one from Pisa's Leaning Tower, and found that they "fell evenly", meaning they hit the ground at the same time. This is likely a myth, as is another falling body – an apple dropping before Isaac Newton – but both led to our current understanding of gravity.

Galileo drops balls from Leaning Tower of Pisa.

The Circulatory System

Galen was a doctor who worked at the height of the Roman Empire and remained the leading authority in Western medicine for some 1,500 years. One of his contentions was that blood was constantly being produced by the liver, pumped around by the heart, and used up by the wider body. In 1628, the English doctor William Harvey presented an alternative take on the way blood moved round the body. He calculated that Galen's system meant that an adult man needed to make 250 kg (550 lbs) of fresh blood a day! Harvey's research, which included treating soldiers wounded in battle and live dissections of animals, confirmed what others had already suggested: the human blood supply was made of loops of vessels – veins brought blood to the heart and arteries carried it away. The smaller, right loop took blood from the heart to the lungs to be replenished with air, and once back in the heart travelled through the left loop, which circumnavigated the body. This set-up is now called the double circulatory system.

William Harvey dissecting a human body.

Science the world over relies on a process of discovery called the scientific method. Follow that method and you will discover truth, or at least you will find out things that are currently impossible to falsify, which adds up to the same thing. The modern take on the scientific method is to observe a system and wonder how it might work based on what already is known to be true. Ask a question about the system, the answer to which is unknown, and then using accepted knowledge come up with a working theory or hypothesis as to what the answer is. That hypothesis allows you to predict the outcome of an event, and you can test this prediction using an experiment.

If the results of the experiment do not bear out your prediction then your hypothesis is false. This is no failure; after all, you've learned something. If the experiment supports the prediction, then your hypothesis is deemed to be a true explanation and so can be used to help figure out something new.

Robert Boyle (left) in his laboratory with Denis Papin.

BACONIAN METHODS

This is a refined version of instructions first formulated by the English statesmen Francis Bacon in 1620. Bacon's version was the *Novum Organum Scientificum* or "New Instrument of Science", which was meant as a guide to thinking more clearly. Bacon hoped that it would be part of a larger work called the Great Instauration. His plan was that this six-part compendium of all knowledge would install a thirst for knowledge that was needed to find fixes for the many problems of the world. Unfortunately Bacon did not live long enough to complete the work, but he certainly had an impact on the advancement of knowledge. The Baconian method, coupled with the stringent empiricism of Galileo, led to an incredible period of discovery in the 17th century. This Scientific Revolution set up the modern fields of science that we rely on today to feed our understanding and drive our technology.

Sir Francis Bacon.

1630 Santorio Santorio, an Italian physician, weighs himself, his food and faeces every day for 30 years to show that some of the material in food is being absorbed by his body.

1638 Galileo publishes his law of falling bodies as part of a compendium of his life's work. The law states that the distance an object travels is proportional to the square of the time spent in motion. This forms a basis of future laws of motion.

1639 English astronomer Jeremiah Horrocks observes the transit of Venus, when the planet moves in front of the Sun, having used Kepler's Laws to predict its timing.

1642 Evangelista Torricelli, Galileo's assistant, builds what becomes the first mercury barometer, a device for measuring air pressure.

1648 Blaise Pascal orchestrates an experiment in France's Massif Central mountains using barometers. These prove the existence of air pressure and show that it reduces with altitude.

1650 Otto Von Guericke invents a vacuum pump that is able to create a near vacuum inside sealed flasks.

1654 Von Guericke uses his pump to evacuate the air from inside two iron hemispheres. He then demonstrates that the air pressure pushing on the outside of the sphere is strong enough to keep them together even when pulled apart by two teams of horses.

• James Ussher, an Irish bishop, calculates the age of Earth using Biblical histories. He states the Universe was created in 4004 BCE.

1655 Dutch scientist Christiaan Huygens proposes that the unusual shape of Saturn observed through telescopes is due to a ring system around the planet.

1656 Huygens invents the pendulum clock, which uses the constant and known swing of a pendulum to keep time.

1660 Robert Hooke presents a law of elastic extensions, now known as Hooke's law. The law describes the relationship between a force and how much a material stretches.

• Robert Boyle uses an air pump built for him by Hooke to investigate the properties of air.

ROBERT HOOKE

Robert Boyle's first assistant was a talented researcher called Robert Hooke, who was on scene for many of the great breakthroughs of the Scientific Revolution, but usually in the background. Nevertheless, Hooke has two discoveries to his name alone: Hooke's law, which relates how the stretching of a material relates to the force applied to it. This law has far-reaching consequences for understanding bouncing, or more correctly oscillations, which are present in everything from chemical bonds to earthquakes. The second discovery comes from Hooke's work with microscopes, a telescope-like device that makes things look bigger than they are. Hooke peered at all kinds of plant and animal specimens through his home-made device. He saw that a sliver of cork was made up of many small spaces, which he likened to the compact living quarters of a monk, known in those days as a cell. Today we still use this term to describe the building blocks of living bodies.

Robert Hooke's cork cells.

NATURE ABHORS A VACUUM

Early developments in physics and chemistry concerned the air. Such research grew out of a problem that had flummoxed even Galileo. In his role as a government scientist he had been asked to help fix a problem with siphons. Siphoning is a clever engineering trick where water (or any liquid) can be made to flow up and over an obstacle so long as the exit is below the entry point. Engineers had used siphons to move water, but they were limited to lifting water about 10 m (33 ft). Engineers wrote to Galileo to ask why.

Galileo used Aristotle to explain the issue. One of Aristotle's many famous aphorisms is "nature abhors a vacuum". If air or anything else were removed from a space, then more air, or in this case water, would always rush in to fill the emptiness. So according to this logic an actual vacuum was impossible, but the mere possibility of one created a pulling force, and it was this pull that drew water through a siphon. Galileo's answer was that the pull of the vacuum was simply too weak to lift water above 10 m (33 ft).

Limits on the ability to siphon water led to investigations into vacuums and pressure.

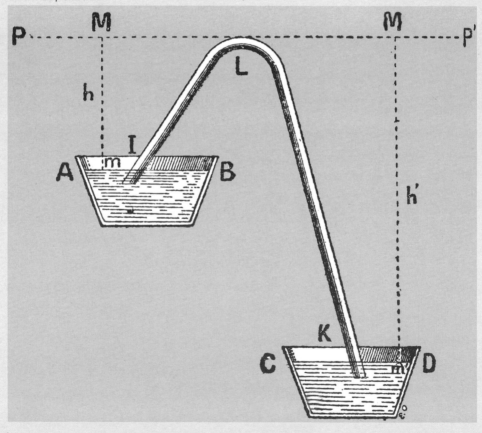

UNDER PRESSURE

After Galileo's death in 1642, his assistant Evangelista Torricelli took up the challenge of learning more. Instead of building a 10-m (33-ft) water column, he constructed a miniature model using mercury, a liquid that was is heavier – 14 times denser – than water. The most enduring of Torricelli's experiments was to fill a glass tube, sealed at one end, with mercury. The open end of the tube was then placed in a bath of mercury. The system was open at one end, but a column of mercury remained in the vertical tube. No matter the initial height of the tube, the mercury always fell to around 76 cm (30 in) – 14 times smaller than a 10 m (33 ft) water column.

The space above the mercury column appeared completely empty. Maybe this space was a vacuum. If so, was the vacuum pulling the mercury up the tube? Torricelli noticed that the height of the

1661 Robert Boyle publishes *The Sceptical Chymist*. This book debunks the myths and magical thinking of alchemy and is seen as the first chemistry text book.

1662 Robert Boyle and Denis Papin use their experiments with air to formulate the first gas law (Boyle's law): gas pressure is inversely proportional to its volume. Two more gas laws follow in the next 150 years.

1663 Otto von Guericke invents an electrostatic generator, which uses a spinning sulphur ball to generate a static electric charge.

1665 Robert Hooke uses the word "cell" to describe the tiny units he sees in plant tissue when viewed through a microscope.

1666 Dane Nils Steensen (or Nicolas Steno) identifies fossilized sharks teeth and thus founds the science of palaeontology, the study of fossils.

1668 Isaac Newton presents his designs of a reflecting telescope to the Royal Society of London.

1669 Phosphorus is isolated by Hennig Brand, who becomes the first known discoverer of an element in history.

DISCOVERING PHOSPHORUS

In 1669, just as Robert Boyle was putting an end to alchemy, a German wizard was sure he had discovered the philosopher's stone. Hennig Brand's raw material was a vat of putrid urine, which he boiled into a syrup, separated out in solid components, which he roasted for many hours. The result was a white solid that glowed in the dark. Brand named it phosphorus mirabilis, meaning the "miraculous bearer of light." Phosphorus proved not to be the magical material Brand had hoped for, but he had discovered a new element, and is the first person in history who is recorded to have done so.

Hennig Brand discovers phosphorus.

mercury fluctuated hour by hour, and this brought to mind the behaviour of a thermoscope, which was a primitive forebear of the thermometer. At first glance, the thermoscope and Torricelli's apparatus appear very similar. However, the thermoscope's tube contains an air bubble that is trapped inside by a column of water, and unlike the mercury tube, the thermoscope is sealed at both ends. Nevertheless, the water column inside the device does move up and down, indicating changes in temperature – or the

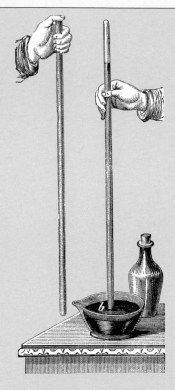

temper of the air, as it was described back then. Warm air has a great temper, so it pushes the water column down making the air bubble grow. Cool conditions sooth the air's temper and the bubble shrinks, allowing the water to rise up.

Torricelli's experimental set-up of a mercury barometer.

WEIGHING THE AIR

All this suggested that the rise and fall of the mercury in Torricelli's tube may be caused by the air, only this time it was on the outside of the system. Today we call a device that responds to the changes in air conditions a barometer. Specifically, a barometer measures air pressure and is a useful tool in forecasting the weather. However, the idea of air pressure was still to be formulated, and sadly Torricelli died before he could do it. Instead the task fell to Blaise Pascal, a French mathematician, who realized that the Torricellian tube could be used to weigh the air.

Pascal's idea required a mountain and a good deal of exertion, and the Paris-based scientist was too sickly for that. In 1648 he convinced his brother-in-law Florin Périer, who lived in Clermont-Ferrand on the edge of the Massif Central mountains, to perform the experiment on his behalf. Périer was supplied with two mercury barometers. The first was set up outside the monastery in the town, and a monk was stationed to check any fluctuations in the column's height (it did not move appreciably). Périer set up the second tube beside the first and confirmed that they both showed the same height. He then dismantled the apparatus and hauled it up Puy de Dôme, an extinct volcano nearby. On the way up he and his team paused and reassembled the device several times until they finally set it up on the peak. As he climbed higher, each time the mercury column fell to a lower level.

Back in Paris, Pascal was delighted with the results because it confirmed this theory. The mercury column was not pulled up the tube by the vacuum. Instead it was forced up by the weight of the air above pushing down on the surface of the mercury bath. At sea level the weight of the air is greatest, and as you go up higher, the amount of air above is reduced. As a result its downward force

Blaise Pascal.

is smaller, and the mercury is pushed up to a lower position.

What Pascal characterized as the weight of the air is now better described as air pressure. Pressure can be applied by any material on any surface, and in modern physics it is defined as the amount of force applied to a unit of area and its unit is the pascal (Pa) in honour of its discoverer. (Air pressure is also measured in millibars, especially by weather forecasters; 1 millibar equals 100 Pa.) A column of water that is 10 m (33 ft) tall exerts the same pressure as the whole atmosphere rising hundreds of kilometres above our heads, and so atmospheric pressure is not strong enough to push water higher than 10 m (33 ft). A machine, or engine, that used high pressures to create force might be able to though...

Florin Périer checks the results of Pascal's barometer experiment on the summit of the Puy de Dôme, France.

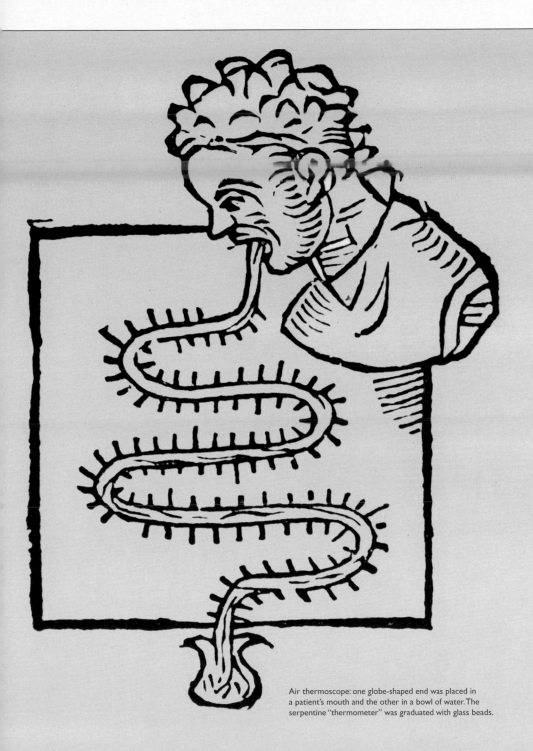

Air thermoscope: one globe-shaped end was placed in
a patient's mouth and the other in a bowl of water. The
serpentine "thermometer" was graduated with glass beads.

THE POWER OF NOTHING

Two years later, a German inventor called Otto von Guericke invented an air pump that could suck out air to create vacuums – or at least very low-pressure spaces. He demonstrated it using a device called the Magdeburg hemispheres, named for the city where they were first used. The cast iron hemispheres fitted together very tightly, but were not locked together in any way. Von Guericke pumped out the air from inside and showed that the pressure of the air on the outside was enough to keep the two sides firmly together. His most memorable demonstration was made for Frederick William I, the Duke of Prussia, where two teams of eight horses were unable to pull the spheres apart.

The Magdeburg hemispheres experiment, which demonstrated the strength of atmospheric pressure.

So it was clear that air was a substance of great strength, but what was it made of? One of the first people to investigate the properties of air was the Irish scientist Robert Boyle. Boyle was a faithful acolyte of Francis Bacon and was blessed with a sizeable fortune that allowed him to build an array of glassware devices for his tests. To help him, Boyle reverse engineered a version of von Guericke's pump (in fact, he had his assistant Robert Hooke do it for him) and named the result his "pneumatical engine". In 1660 he published his first book, *New Experiments Physico-Mechanicall, Touching the Spring of the Air and Its Effects*, which related such things as how a feather falls as fast as a stone when the air is removed from around it. Also sound does not pass through a jar where the air has been removed, nor do candles burn nor animals live for long.

Much of this sounds obvious and trite, but Boyle was the first to observe and record it in a systematic way. His next book, published in 1661, was called *The Sceptical Chymist*, in which he debunked the magical thinking of the alchemists and set out how one could investigate substances in a scientific way. In so doing he more or less founded chymistry – soon to be chemistry.

Boyle's work on air proved that it was made of some kind of invisible material, but to him it was one substance: air. The idea that there could be several types of air, or gases, was only just emerging. Nevertheless, Boyle contributed what is now known as the first gas law, or Boyle's law, which states that the pressure of a gas (air) is inversely proportional to its volume. In other words, squeeze a gas into half the volume and its pressure will double (as long as the temperature stays the same). More gas laws were to be added over the following centuries, and they would be the keys that unlocked the atomic nature of matter.

Robert Boyle.

Sir Isaac Newton is the superhero of the Scientific Revolution, the behemoth that bestrode research in physics for the next 200 years. His laws of motion and gravity created a clockwork universe that made perfect sense (for a while, at least), and the mathematical techniques he developed made it possible to quantify natural phenomena, even those that are forever changing and never static, in meaningful ways.

Despite his place in science history, Newton was something of an anachronism. While his contemporaries were building apparatus and performing experiments, Newton sat around thinking, like some kind of latter-day Greek sage. His most famous observation was an apple falling from a tree. In addition, Newton was notoriously secret, especially as a young

Isaac Newton observing an apple falling from the tree.

man, and jealous and cruel in middle age as he sought to protect his claims to his momentous discoveries.

Isaac Newton.

1671 Giovanni Domenico Cassini, an Italian astronomer, identifies Iapetus, the first moon of Saturn to be discovered.

1674 Dutch textile merchant Antonie van Leeuwenhoek develops a simple microscope to study threads up close. He also uses the device to discover tiny singled-celled protozoa which he called animalcules. These are the first observations in the field of microbiology.

1675 The Royal Observatory is founded at Greenwich near the centre of London as a research base for the king's astronomer. The observatory becomes the position of the prime meridian from which the world's time is now measured as GMT or Greenwich Mean Time.
• Gottfried Leibniz publicizes his technique of infinitesimal calculus, and begins a long-running dispute with Isaac Newton over who developed this new mathematical method.

1676 Danish astronomer Ole Rømer attempts to find a figure for the speed of light using delays in the expected and actual appearance of Jupiter's moons. His measurement is the first of its kind but is 25 per cent too slow.

1678 Dutch scientist Christiaan Huygens proposes that light beams behave like a wave. This contradicts Isaac Newton's theory that light is made of tiny particles, which move according to his laws of motion.

1682 The English researcher Nehemiah Grew publishes the most detailed anatomy of plant structures to date, plus descriptions of their functions, in his book *Anatomy of Plants*.

1687 More than 20 years after he completes his research, Isaac Newton presents his universal law of gravitation and laws of motion in *Philosophiæ Naturalis Principia Mathematica* (*Mathematical Principles of Natural Philosophy*).

1698 Thomas Savery builds a primitive form of the steam engine for pumping water from a mine. He calls the device The Miner's Friend; or, An Engine to Raise Water by Fire.

SPEED OF LIGHT

Galileo is reported to have tried to measure the speed of light by timing the delay from lanterns a large distance apart. Obviously this did not work – the beams travelled so fast that their appearance was effectively instantaneous. In 1676, the Danish astronomer Ole Rømer had a better idea. Instead of light from a lantern, he was going to time how long it took for light to reach him from Io, Jupiter's first moon. Kepler's laws told him exactly when it would move out from behind Jupiter. However, he did not see it until 10 minutes after that. This gave him the data to calculate the speed of light at 220,000 km (136,702 miles) a second. This was 25 per cent slower than the accepted value, but was the first scientifically sound measurement.

Ole Rømer using a telescope.

NEWTON'S LAWS

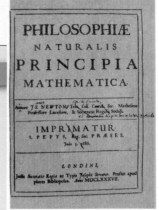

SECRET BEGINNINGS

The bulk of the discoveries for which he became most famous – his theory of universal gravitation and laws of motion – were made in the mid 1660s when Newton, still in his early twenties, was living in the seclusion of his family farm in Lincolnshire. He had left his post at Cambridge University to avoid the Great Plague that was sweeping the nation. However, he did not publish his

Isaac Newton's *Philosophiæ Naturalis Principia Mathematica* was published in 1687.

PUBLISHED RESULTS

For example, in 1684, Robert Hooke, now an established member of the scientific community and a leading figure in the Royal Society, England's academy of science based in London, was discussing his ideas about how the force that governed planetary motion works with the astronomer Edmond Halley and architect Christopher Wren. (Halley is now famed for his work predicting the return of a comet every 76 years, a comet that now bears his name, while Wren was England's foremost builder responsible for the great dome of St Paul's Cathedral.) Hook was proposing that force obeyed an inverse square law and claimed to have the mathematics to back it up. Such a relationship, which states

Newton's law of universal gravitation.

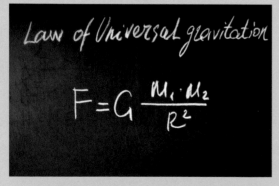

that an effect diminishes in proportion to the square of distance, had been discussed for a range of phenomena since the 14th century. Hooke's suggestion was that the pull of gravity between objects 1 mile (or any unit of distance) apart, was four times as great as the force between them when they were 2 miles apart. (Dividing the distance by two increases the force by a factor if two squared, which is four.)

Wren and Halley were unconvinced by Hooke's partial proof, and asked Newton to check. Newton said that he already knew all about it but had mislaid his papers. With Halley's sponsorship, Newton decided to publish his findings on the mechanics of motion on Earth and in the heavens. When Hooke read *Principia* in 1687, he accused Newton of stealing his ideas (not for the first time). Newton declared his work predated Hooke's lesser attempts (and historians agree, saying that any similarities were arrived at independently.) Newton famously retorted that "if I have seen further, it is by standing on the shoulders of giants." This was a wicked barb because it suggested that Hooke was not one of the titanic minds who had contributed to Newton's thinking. In addition the rebuke was also targeting Hooke's short stature, with several reports saying he was a hunchback.

results for more than 20 years with the eventual release of *Philosophiæ Naturalis Principia Mathematica*, fondly remembered by the more numerate among us as simply *Principia*, in 1687.

History often reflects that this book was a revelation of fresh thinking. It was without a doubt a work of astonishing genius, but we should not forget that other people had a hand in its development.

Some of these contributions came before Newton had turned his attention to the subjects of motion and gravity, others came during the interim period when he kept his work secret. Those who saw similarities between their thinking and Newton's work and dared to suggest that they shared some of the credit for it were dealt with harshly.

A UNIVERSAL LAW

Newton was happy to admit inspiration from the likes of Aristotle, Kepler, and Galileo. For example, Aristotle asserted that objects with more "gravitas" – meaning heft or weightiness – fell to the ground faster than lighter ones. Galileo had shown that, contrary to this assertion, objects of all sizes fell "evenly," meaning they covered the same distances in the same time.

Galileo also showed that projectiles, objects thrown through the air, follow a parabolic path. Around the same time, Kepler produced the mathematical description of the way planets moved in orbit, showing they followed elliptical paths. Kepler believed that they were held in place by some effect of magnetism. It was Newton who connected all these ideas together to create a universal law of gravitation. As is well documented, he is said to have had his epiphany after watching an apple fall from a tree. In that moment he realized that the force of gravity pulling objects to Earth was the same force that held the Moon and planets in their orbits.

The universal law of gravitation states that gravity is always an attractive force, never repulsive. All objects with mass produce gravity which pulls them together. The force of gravity is proportional to the masses of the objects involved. It is also inversely proportional to the square of the distances between them. So in summary, the pull of gravity from a heavy object like Earth is much larger than the pull of gravity from an apple. As a result, objects are pulled toward Earth and not toward the apple. However, Newton's explanation of gravity does show that the Earth does move upward to meet the plummeting apple, but by such a minute amount that is utterly negligible. Secondly, as the distance between objects increases, the pull of gravity decreases rapidly.

Newton's cannonball.

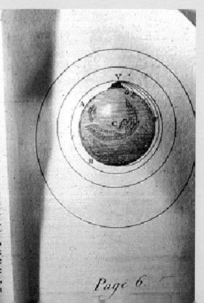

OF THE SYSTEM

miles before it arrived at the till at last exceeding the limits of earth, it should pass quite by without touching it.

A F B represent the surface of the ..., C its center, V D, V E, V F, the lines which a body would de- ... if projected in an horizontal di- ... from the top of an high moun- ... successively with more and more ... ty. And, because the celestial mo- ... are scarcely retarded by the little resistance of the spaces in which ... are performed ; to keep up the pa- ... of cases let us suppose either that ... is no air about the Earth, or at ... that it is endowed with little or no ... er of resisting. And for the same ... on that the body projected with a velocity, describes the lesser arc V D, ... with a greater velocity, the greater VE, and augmenting the velocity, ... es farther and farther to F and G; ... he velocity was still more and more ... gmented, it would reach at last quite ... yond the circumference of the Earth, ... return to the mountain from which ... was projected.

Page 6.

73

FORCE AND MOTION

Like all mathematicians, Newton understood that parabolas and ellipses were shapes governed by similar rules. His great breakthrough was to see that the parabolic path of a projectile like a cannonball was produced by gravity in the same way as the ellipse of a planet's orbit. The difference was speed. If a cannonball could be fired high and fast enough, its curved path would extend all the way around the planet and start where it left off. Move any slower and the ball would inevitably fall to Earth, and if propelled to even greater speed, its path would escape the influence of Earth's gravity altogether.

Of course, Newton's description of how things moved did not end with gravity. One confusion is that if larger objects are creating greater gravitational forces (in keeping with Newton's law) then why do they not move faster than smaller ones as Aristotle had believed. Newton's Laws of Motion (there are three) generalized the subject to explain how any kind of force – gravity, magnetism or just a push – worked to alter the motion of any object.

Naming the Rainbow

After completing his work on gravity and motion to his satisfaction, Newton moved on to other fields, including the study of light. In 1668 he presented his design of telescope to the Royal Society, which used a curved mirror to collect light and focus images for magnification. Such a device was much cheaper to build than one that used only lenses, and today the largest telescopes of all, including Hubble, follow a similar design. We also have something else to thank Newton for – the colour spectrum that makes up our rainbow. Newton split white sunlight using a prism and divided the full range of colours – the spectrum – into easy-to-understand shades. Six of them were obvious: red, yellow, orange, green, blue and violet. However, he felt a seventh would be more auspicious. (Despite being remembered as a man of science, Newton was also one of the last alchemists with a great interest in magic and the supernatural.) To round up the spectrum to seven, Newton invented a new kind of deep blue, indigo, which was named after a natural dye hailing from India.

Newton working with refracted light.

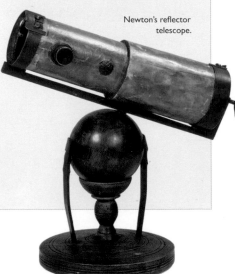

Newton's reflector telescope.

THE CALCULUS

Newton is one of the inventors of calculus, a mathematical technique that breaks a constantly changing value into an effectively infinite set of fixed values. In so doing it makes it possible to analyze the behaviours of natural phenomena such as waves, sound and growth. As usual, Newton guarded his methods, devised in the mid 1660s, jealously and only made them public when Gottfried Leibniz published something similar in the 1670s. Newton accused Leibniz of plagiarism, and began a long and bitter dispute that spilled over into wider attitudes between scientists working in Britain and continental Europe. Today it is clear that both men developed calculus independently. Leibniz's version is easier to use and his notation eventually supplanted Newton's.

Gottfried Leibniz.

THREE LAWS

The first law says that an object will maintain its state of motion until a force acts on it. This obviously means things stay still until forced to move, but it also illustrates how moving objects will maintain their speed and direction in perpetuity until a force is applied. This first law is describing inertia, which is a resistance to change in motion exhibited by all masses.

The second law says that the acceleration of a body is proportional to the force applied to it. But it goes further, using the simple formula Force (F) = Mass (m) x Acceleration (a). So a given force will accelerate a light object more than a heavier one. To get them moving at the same speed, you'll need to apply a greater force to the bigger object. That is one way of understanding why gravity makes all objects fall at the same speed – it pulls harder on the bigger ones but they also resist moving more.

This is all pretty intuitive stuff so far, but the third law is less so. It says that for every action there is an equal and opposite reaction, which means when you push an object with a force (the action) that object will push back with an equal force directed in the opposite direction. This is inertia in action and is the reason why force creates motion in the first place. Without it, forces would have nothing to push against. Newton's description of motion and force works well enough to send spacecraft into orbit and fly all the way to the Moon. However, its clockwork nature did not fit with the behaviours of light and other phenomena that have no mass. To solve that mystery would take Albert Einstein, perhaps the only physicist to outdo Newton.

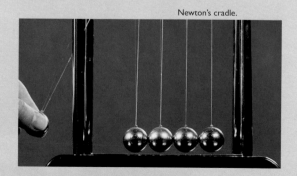

Newton's cradle.

The power of the scientific method was well illustrated by the surge in fundamental discoveries in the 17th century, spearheaded by the likes of Galileo, Pascal and Newton. As the century drew to its conclusion, another signal that science was on the right track was an invention called the Miner's Friend; or An Engine to Raise Water by Fire. This was the first practical steam engine, showing that science is not just a way of finding out the truth about nature, it also results in useful inventions that solve problems by taking advantage of the laws of nature as they are revealed.

The engine's English inventor, Thomas Savery, had been inspired by the work on air and pressure by Robert Boyle and especially his right-hand man Denis Papin. Frenchman Papin was an integral figure in Boyle's investigations into the link between air pressure and volume. One of his claims to fame was the invention of the pressure cooker in 1679, where he demonstrated that the effects of temperature on food were multiplied when the process took place under high pressure. In 1690 Papin repurposed his cooker design to harness the mechanical power of the expanding hot air inside and ended up building a simple piston engine.

Denis Papin.

An old steam locomotive.

MINER'S FRIEND

Savery's "engine" was a pump, initially designed to remove water from mines but later to find a use (for a short while, at least) in the plumbing systems of grand houses. It did not use a piston and had no moving parts at all. Instead the pump worked by manipulating the pressure inside the parts of the device and thus allowed the air pressure outside to push water up through pipes. The process began with a boiler releasing hot steam into a large chamber, or "working vessel", and from there into a pipe that reached down into the water to be lifted. When this pipe was full, the steam supply was shut off using a tap. None of this was automatic; a human

1701 Jethro Tull invents a horse-drawn seed drill which makes it much easier and faster to plant seeds, thus helping to reduce the number of people needed to work on farms.

1702 Antonie van Leeuwenhoek, the inventor of the microscope, describes Vorticella, which is the first known ciliate. Ciliates are single-celled organisms covered in hairlike extensions called cilia.

1703 Gottfried Leibniz develops the binary number system that counts using only two numerals, 1 and 0. These numbers or digits will become the basis for the inputs and outputs of all computers.

1704 Newton publishes his theory that light is made from streams of corpuscles, or particles. This is at odds with the wave theory already proposed by Huygens.

FAHRENHEIT SCALE

In 1724, a Dutch instrument-maker invented a temperature scale that still retains his name. Daniel Fahrenheit was an expert instrument-maker and his success lay in the way he perfected the specifications of mercury thermometers so they could always be relied upon to give the same readings. Zero degrees (0°) Fahrenheit was set using a mixture of water and salts, which was "frigorific", meaning it always chilled to the same cold temperature when mixed. The upper limit was set at 96°, which is the temperature of the human body.

Daniel Fahrenheit
with thermometer.

operator controlled the process. The hot steam in the vessel and pipe began to cool, aided by cold water splashed on the outside. The cooling steam condensed into water, reducing the pressure inside the vessel to a partial vacuum. Thus the air pressure pushing down on the water outside the engine forced water up the pipe into the working vessel. The steam in the boiler was then let back into the vessel, forcing the water inside up and out of the system. Savery's engine took the heat energy released from a fire (in the boiler) and used it to lift up water. That is quite something.

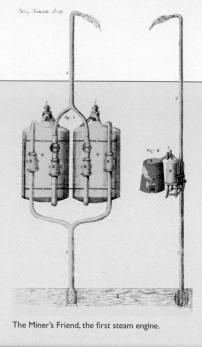

The Miner's Friend, the first steam engine.

ATMOSPHERIC ENGINE

However, Savery's engine was convoluted to use, slow, low-powered, and prone to frequent failure when the cast iron vessels cracked. Nevertheless, he was awarded ownership rights over any fire-powered water pump. That meant that the next innovator, Thomas Newcomen, was forced to go into partnership with Savery. Newcomen's "atmospheric engine" was more effective and reliable than Savery's. The first one was built at a copper mine in Cornwall in 1712, and unlike with the Miner's Friend, its utility was quickly recognized by mine owners across Britain.

Newcomen's design did owe something to Savery's system, but it also owed much to the piston system devised by Papin. Papin's designs had moved on in the ensuing years, and he had devised a steam-powered paddle boat and a pump system remarkably similar to Newcomen's. Fortunately (for Newcomen at least) by 1712 Papin had died in poverty and obscurity so there could be no disputes over ownership...

Newcomen's atmospheric engine.

The atmospheric engine directed high-pressure steam into a sealed cylinder and that forced a piston to rise upward. That motion was translated as plunging the pumping gear into water. Then cold water was sprayed into the cylinder, making the steam within condense very rapidly thus reducing the pressure. Air pressure pushing down on the piston forced it to move down inside the cylinder – and the pump was raised up (shifting some water in the process). This downward motion was the power stroke of the engine that did the hard work of lifting the pump and water.

1705 Edmond Halley uses the observational records of a comet to calculate its orbital period, predicting it will return every 76 years. He is correct and the comet is named after him.

1712 Thomas Newcomen invents the atmospheric engine which forms the basis of the next 50 years of steam engine designs.

1724 Daniel Gabriel Fahrenheit, a German physicist, develops a new temperature scale based on the freezing point of a salt and water mixture and human body temperature.

1729 Pier Antonio Micheli discovers that fungi have significant differences to plants and should be classified as a different kind of organism.

WATT ENGINE

The atmospheric engine became the mechanical workhorse of the early Industrial Revolution. As well as being used as a pump, it could create rotary motion via a crankshaft. However, its slow and ponderous action was ill suited to that task. In the 1760s James Watt, the most famous steam engineer of all, would redesign the engine cylinder so steam could act on either side of the piston, forcing it down as well as up. That innovation allowed engines to work much faster and with enough power to drive wheels and other machines.

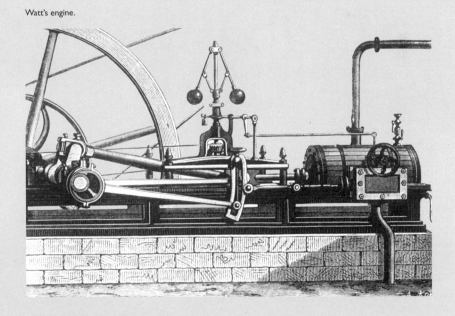
Watt's engine.

The air pump invented by Otto von Guericke in the mid 1650s was the seed that sprouted many important investigations — which eventually blossomed into the modern science of chemistry (we'll see how a bit later on.) However, the inveterate German innovator also kick-started the scientific study of electricity with his invention of the sulphur globe in 1663. This device harnessed the way friction between two substances could create mysterious sparks and electric shocks, a phenomenon that Thales had known almost 2,000 years before. In so doing, Von Guericke invented the first electrostatic generator. The globe of solid sulphur was spun by hand, and the friction between the two resulted in electricity building up on the device. The globe exhibited attractive and repulsive forces, and could be discharged as a visible spark. In those days, these behaviours were seen as evidence of some "electrical virtue" being released from an object.

Hauksbee's electrical machine.

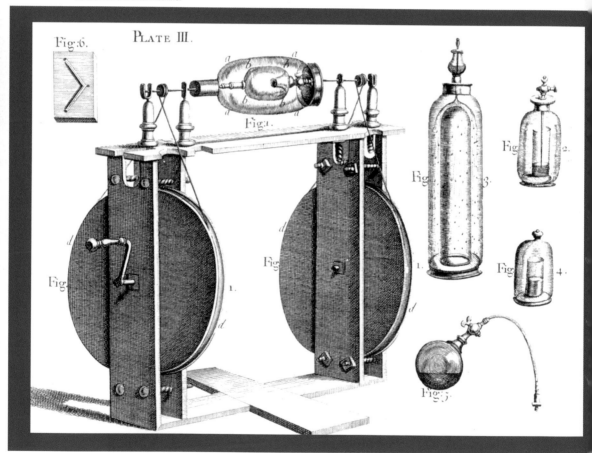

1730 Stephen Gray investigates the abilities of materials to conduct electric charge, or insulate against it. This works culminates in the "Flying Boy" experiment.

1732 Cobalt is discovered by Georg Brandt as he ˌˌˌˌˡˡˡˡˡˡˡˡˡ ˡˡˡ ˡˡˌ source of the blue colour in glassware and porcelains.

1735 Antonio de Ulloa discovers a new metal, later named platinum, which is derived from the Spanish term for "little silver".
• Carl Linnaeus's *Systema Naturae* is published, setting out a way of classifying living things by arranging them in a hierarchy of groups.

1737 William Champion designs a technique for producing pure zinc on a large scale, using it to make brass, an alloy of zinc and copper.

1739 The French Geodesic Mission sends missions to Ecuador and Lapland to measure the shape of Earth. It is found that the planet is flatter at the poles and bulges around the equator.

1742 Anders Celsius, a Swedish astronomer who took part in the French Geodesic Mission to the Arctic, devises the first centigrade temperature scale that is based on the freezing and boiling points of water.

1745 Pieter van Musschenbroek is credited with the development of the Leyden jar, a device for storing electric charge.

1746 Jean-Antoine Nollet sends an electric charge through 200 monks to measure the speed at which electricity moves.

CELSIUS

In 1742, Anders Celsius, a Swedish astronomer, developed a new temperature scale based on the melting and boiling point of water. Being an Arctic explorer, Celsius was interested in freezing temperatures, so he set 0° at water's boiling point and 100° at the freezing point. The simplicity of such a "centigrade" scale was obvious when compared to the slightly confused Fahrenheit system, and the values were soon swapped so 0° was reset as freezing point. Carl Linnaeus insisted he had developed such a system before Celsius and that it should be °L not °C. The world of science paid no heed.

Anders Celsius is commemorated in a Swedish stamp..

HAUKSBEE'S DEVICE

Von Guericke's electrostatic generator caught the attention of scientists across Europe. In the early 1700s, a young instrument-maker working at the Royal Society of London began to investigate the merits of merging the functions of both von Guericke's inventions. Following a suggestion of Isaac Newton (who had just taken over the stewardship of the Society), Francis Hauksbee used an air pump to evacuate the air from inside a glass sphere and used that in place of a ball of sulphur in an electrostatic generator. Instead of spinning the device by hand, he cranked it with a handle, but he found he still needed to touch the turning sphere to create the friction needed to generate a charge.

Hauksbee demonstrated his device to the gathered fellows of the Society, insisting that all light was extinguished from the meeting chamber. In the dark, the electrified globe was seen to glow in the dark. Hauksbee revealed that the effect was due to a tiny amount of mercury inside the glass. (He had discovered the mechanism behind gas discharge lamps, where a diffuse vapour releases light when a powerful electric current passes through them. This effect is behind what used to be called neon lights and is used in the "low-energy" light bulbs used in homes today.)

ELECTRICAL DISPLAY

Hauksbee's device was much more powerful than von Guericke's – and more fun. Soon a new breed of entertainer, called the "electrician", was putting on shows in the great houses of Europe. The showmen would deploy electrostatic forces in various ways, set fire to things with sparks and even offered "electric kisses" where a female volunteer was charged up with a Hauksbee device. A suitably gallant man would then approach for a kiss only to find sparks literally flew between the pair's lips!

The electric kiss.

INSULATOR AND CONDUCTORS

In the 1720s, Stephen Gray, a teacher working in London, began to demonstrate "electrical virtue" in lessons. He used a hollow glass tube, rubbing it with a cloth to generate a static charge. He noticed that although he was only rubbing the glass, the corks used to seal the tube at either end were also being filled with electric virtue – and so attracted dust and feathers. The "virtue" must be able to flow from one object to another, Gray reasoned, and he carried out a series of tests. Fresh, oily wood seemed to carry the electricity, as did ivory and metal wires. Gray even sent electricity along a bundle of hemp fibres 24 m (79 ft) long with an ivory ball at one end. Touching the charged glass tube to the other end made the ivory ball come alive with electrical virtue so lightweight objects stuck to it. Gray noticed that the effects disappeared when any of the hemp cable touched the ground, or, as Gray put it, was "earthed." So he hung the cord on silk threads. He categorized silk as an insulator, because it blocked the progress of electricity, while he termed hemp, metal, and ivory as conductors in that they allowed the electricity to pass through.

Gray illustrated his discovery by recreating the hemp experiment in a more flamboyant form, called the "Flying Boy". One of Gray's pupils was hung on a board above the ground using silk cords, and thus completely insulated. Gray used his glass rod to charge the boy's feet, and in response, the boy's hands became charged and were able to pull brass flakes up from the floor without touching them – as if by magic.

Gray ran electricity through a long hemp cord suspended above the ground.

Gray's "Flying Boy" electrical conduction demonstration..

TWO FLUIDS

Of course, Hauksbee, Gray or any contemporary did not use the word "charge". Upon hearing of Gray's work, Charles François de Cisternay du Fay, a French gentleman scientist, took up further research into electricity. He argued that there were two distinct types of electricity, which du Fay described as fluids. Sulphur, amber and similar substances made "resinous" electricity, while glass made a "vitreous" electricity. Objects filled with the same "fluid" would repel each other but those with differing "fluids" were attracted to each other. Hence, "opposites attract and likes repel" rule was getting somewhere nearer to the truth of the matter.

Charles François de Cisternay du Fay.

STORING ELECTRICITY

If electricity was made of fluid, then perhaps it could be stored in a bottle or jar? In 1745, a German scientist called Ewald Georg von Kleist attempted to do just that. He lined the inside of a glass jar with silver foil and added a little water. He then connected the foil to an electrostatic generator with the idea that he would "fill" the bottle with electrical fluid, mixing it in with the water. This did not happen as he expected but when he touched the foil he received a dangerous electric shock that threw him across the room. He survived, and just as importantly, his jar was a means of storing electricity.

Shortly afterward, the Dutchman Pieter van Musschenbroek, an inventor and an acquaintance of von Kleist, built a similar device and showed it to scientists at the University of Leyden (spelled Leiden by the Dutch). Henceforth, the apparatus became known as the Leyden jar. Today we understand it as being a capacitor, a device for storing electric charge by

Leyden jar.

keeping two large conductors slightly separated. (These are very commonly used in everyday electrical devices. For example, touchscreens rely on capacitor technology.) The final design of a Leyden jar used three containers, two metal ones (conductors) and one glass one (an insulator). The smaller metal container slotted inside the glass one, which in turn sat snugly inside the larger metal vessel.

The inner metal container received the "electric fluid" from the generator, but that only worked if the outer metal one was earthed – often by being held by the researcher. When the inner and outer containers were connected, the electricity flooded out as a spark.

Du Fay's two-fluid theory did not adequately explain what the Leyden jar was doing. The necessity to earth the outside of the jar meant that electricity was moving in or out of the outer conducting layer. How could vitreous electricity flow inside while resinous electricity filled the outside? The American researcher Benjamin Franklin is most famous for proposing an experiment to charge a Leyden jar with a bolt of lightning using a kite to carry a metal wire into the thundercloud. This is exquisitely dangerous; it is more than likely he never did it, and those that did frequently died in the process. However, his more significant contribution was to propose a one-fluid theory of electricity. His idea was to explain electricity as a lack or an excess of a single substance. He reclassified vitreous electricity as an object being positively charged (as in filled) with electrical fluid, and resinous electricity as it being emptied of the fluid, or negatively charged. Today electrical charge is one of the fundamental properties of matter, and its investigation is instrumental in our understanding of chemical bonds, light, and subatomic particles.

Benjamin Franklin.

CLASSIFYING LIFE

As well as setting the scene for the science of physics for at least a millennium, Aristotle was also the founding figure of biology. His most significant contributions were to the field of taxonomy, which seeks to organize the great diversity of life on Earth into different groups.

FINDING TYPES

As the son of a physician, Aristotle would have been well versed in human anatomical features, their form and function, and during his travels he turned this way of understanding to other life forms. His natural inclination was to divide the living world according to its anatomical and functional differences, and so he split it in two, into plants and animals. Animals were further divided into those that had blood and those that did not – not the red stuff that Aristotle was looking for, anyway. He then grouped animals that looked the same into groups which he called genera (the singular form is genus). This word broadly translates as "type" but more specifically means "of the same origin"; we get the words genre, gender, generic, genesis and genetic from the same root. The modern form of taxonomy, which has its genesis in the 1750s, also makes use of Aristotle's genus grouping although in a much more highly defined way.

GENERA

In Aristotle's system, all red-blooded animals were split into five genera. The first contained four-legged animals that give birth to their young, which would have included most of what we describe as mammals today. The second genus was birds and the third included the four-legged animals that lay eggs, which bundled the reptiles and amphibians into one group. The fourth and fifth were the fish and whales. Aristotle correctly saw that dolphins and whales were not fish but he could not regard them as mammals, either. (The word dolphin comes from the Greek for "womb fish.")

Dolphins are marine mammals.

1750 Nicolas de Lacaille makes observations from South Africa to create a detailed survey of the southern hemisphere sky.

• Joseph Black isolates fixed air, the first gas to be identified. Today it is known as carbon dioxide.

• Kew Gardens is opened in London, England, as part of the trend for building botanical gardens to study plants over long periods.

1751 Nickel is isolated by Axel Cronstedt. The metal is named after the German word for demon, since its toxic minerals were often mistaken for more valuable copper ores.

1755 Magnesium metal is identified by Joseph Black in his work on magnesia alba (magnesium carbonate). However, the element is not purified until 50 years later by Humphry Davy.

1757 The sextant, a device based roughly on a sixth of a circle, is invented as the latest navigational tool for measuring the angular height of the Sun, which is needed for calculating latitude. The eyepiece uses mirrors so the operator can see the horizon and Sun at the same time.

1762 Joseph Black describes latent heat, a hidden form energy that is taken or released when a substance changes from one state – solid, liquid, or gas – into another.

BLOODLESS ANIMALS

He grouped the "bloodless" animals thus: cephalopods, such as octopuses and cuttlefish; many-legged sea animals, such as crabs and shrimps; many-legged land animals, which comprised all the insects, spiders and other creepy-crawlies; and shelled animals, such as molluscs and sea urchins. (The mouth structure of the sea urchin is still called "Aristotle's lantern" thanks to his original description.) Aristotle's final animal group was the "plant-animals", which includes jellyfish and anemones.

Aristotle viewed nature as a continuum, which began with the lifeless earth and rocks, passed through plants, "plant-animals", and then animals, and ended with the human race. This hierarchical view is a lasting legacy of Aristotle's biology and one that has no basis in science but is still hard to shake off even today.

Sea urchins are related to starfish.

ANATOMICAL SYSTEM

Modern taxonomy arose from the work of Carl Linnaeus. He was a Swede whose clergyman father, a certain Mr Ingemarsson, had restyled himself with a Latinized surname, and as a child Carl spoke mostly Latin at home. Carl became a herbalist and found that he was often confused by the range of names used for one plant. So he devised a fool-proof system for assigning an unambiguous name to living things, which could be understood the world over, and he was going to use Latin to do it. Now known as the Linnaean nomenclature, or binomial naming, the system gives each organism two names. The first is the generic name, referring to the genus, or wider grouping of the organism. For example, the big cats, such as lions and tigers, occupy the

Panthera genus. Each member of a genus has its now specific name, which denotes its species, or particular unique type. So, the lion is *Panthera leo* and the tiger is *Panthera tigris*.

Linnaeus began working on this idea in the 1730s and, like Aristotle before him, grouped organisms according to their anatomical features. He extended his classification into a hierarchy. Each organism belongs to a species, which in turn is part of a genus. Genera are then grouped into families, which fit into an order, then class and then phylum. So by way of illustration, the humble lion, *Panthera leo*, is in the cat family (Felidae), the carnivore order (Carnivora), the mammal class (Mammalia) and vertebrate phylum (Chordata). The final Linnaean

Carl Linnaeus.

group was kingdom, and he identified two: Plantae and Animalia. By 1753 Linnaeus had perfected the system for the plant kingdom, and the animal kingdom was included by 1758 in the 10th edition of his book, *Systema Naturae*.

The Linnaean system of taxons, or classifications, is still used today. The kingdoms have been extended, adding Bacteria, Protista (complex unicellular life-like amoebas), Fungi (Linnaeus thought them plants), and Archaea, which are very primitive, perhaps the most primitive, forms of bacteria-like life. However, modern taxonomists never rely solely on anatomical features to classify a species, because these can be misleading. Instead they prefer to compare the genetic makeup of species to figure out which is related to which. They also look at the way organisms develop as embryos. The precise path they take is the best indicator for how large groups, like the phyla, are related to each other.

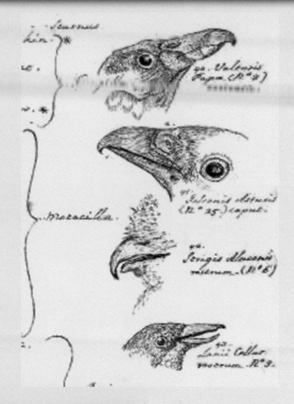

A page from Linnaeus's *Systema Naturae*.

Panthera leo.

Since the dawn of natural philosophy it had been assumed that air was a single substance, one of the elements from which all other materials were made. Robert Boyle's investigations into the "spring of the air" are seen as the foundational event of chemistry, when the hocus pocus of alchemy began to be weeded out by the scientific method. However, the first clue that air was in fact a more complex material than it appeared had been revealed 50 years before by Jan Baptist van Helmont, a mystical alchemist from Flanders. Van Helmont believed that air and water were the only two elements. Fire was not a substance at all, he said, while earth could be turned into water and air (or so he claimed). This stance is due in part to an experiment he did with a willow tree. For five years, van Helmont grew a tree in a barrel, noting the weight of the soil and tree and recording how much water he added over the weeks and months. The tree gained about 70 kg (155 lb), while the soil's weight did not change at all. Van Helmont's conclusion was that plants turned water into wood and other body tissue.

Jan Baptist van Helmont.

1766 Hydrogen gas is isolated by Henry Cavendish. He calls it flammable air but the modern name, which means "water former" was coined later by Antoine Lavoisier.

1770 Abraham Darby III builds a bridge entirely from cast iron. The location in the English Midlands is now called Ironbridge.

• Manganese, a hard metal, is discovered by Torbern Bergman.

1772 Nitrogen is isolated from exhaled air by Daniel Rutherford.

• Carl Wilhelm Scheele suggests that some rocky minerals contain a heavy metal. This is isolated and named barium by Humphry Davy in the 1800s.

1773 Carl Wilhelm Scheele is the first person to isolate oxygen, which he names fire air. Joseph Priestley does the same in 1774 by another method, and is credited with the element discovery because he publishes his finding first. The gas is given its name by Antoine Lavoisier in 1777.

• John Harrison's chronometers are accepted as an effective way to calculate longitude, but there are still concerns over the great cost of these clocks.

LONGITUDE

In 1707, the Royal Navy suffered a terrible disaster as a fleet of its ships foundered on rocks due to a navigational error, and 1,400 sailors died. At the time there was no accurate method for calculating longitude – locations east or west on the globe. A competition was started to encourage innovators to find a solution. The judges assumed an astronomer would solve the issue, and were surprised when a clockmaker called John Harrison presented a method. For every 15 degrees of longitude, the time (as measured at noon when the Sun reached its highest position) varied by one hour. So if you knew the time at a known location, such as London, you could always figure out how many degrees east or west you were. Pendulum clocks did not keep time well on board rolling ships, but Harrison built spring-powered clocks of such precision that they kept time more than well enough. It took until 1773 – after 3 years of work – for Harrison to convince the navy that his plan worked, and he was awarded a cash prize. A clock or chronometer is still a valuable instrument on board any ship.

Harrison's chronometer.

CHAOTIC SPIRIT

When wood was roasted slowly it became charcoal, something that had been perfected before the Bronze Age. Van Helmont noticed that once the charcoal was burned up, its ash weighed less than the original material. He concluded that the missing weight was from a "*spiritis sylvestris*" or a woodland spirit, which escaped from the fuel.

Van Helmont concluded that missing weight from the ashes was due to *spiritis sylvestris*.

FIXED AIR

In 1750, a Scottish medical student called Joseph Black was researching a medicine for kidney stones. He knew that he could attack the little solid lumps with all kinds of chemicals but they were all too damaging to be used as medicine. Instead he chose to investigate the effects of a mild alkali, a white powder known as magnesia alba. It did nothing at all, but Black noted it loosened the bowel somewhat. Nevertheless, Black continued his research into the material, noting that when mixed with oil of vitriol (now called sulphuric acid) it fizzed out bubbles of what he assumed was air. However, when he heated the crystals first, the acid had no effect. Heating had no visible impact on the white powder, but he did find that the chemical was lighter in weight than before. Something invisible had been released. He could recreate the same drop in weight by mixing fresh magnesia alba with acid. The same invisible material was being released by acids as by the heating.

Joseph Black.

Black isolated the air being released, and found that he could recombine it with the residual solids, thus creating magnesia alba again. Black called the mysterious substance "fixed air" because it could be collected by the solid crystals and then released again. Black found that animals breathed out fixed air, it was released by fermenting beer, and it was released by heating other substances such as lime. Fixed air was the same stuff as van Helmont's *spiritis sylvestris*. Today we call it carbon dioxide, but pneumatic chemistry needed a few more breakthroughs to figure that out.

This spirit was an agent of chaos, which van Helmont, a Dutch speaker, spelt as "gahst" – now simplified to gas. This contribution, jumbled and murky though it is, earns van Helmont the position of the first "pneumatic chemist," or an investigator of gas. Pneumatic chemistry was a crucial field in early chemistry – but real progress would have to wait for more than another century.

1774 Carl Wilhelm Scheele produces pure chlorine but only Humphry Davy recognizes it as an element in 1808.

1776 David Bushnell designs Turtle, a one-person mini-submarine designed to place mines on enemy ships. It is powered by hand-cranked propellers for controlling depth and providing thrust.

1777 Antoine Lavoisier discovers that humans and other animals take in oxygen and give out carbon dioxide. This confirms the theory that life uses oxygen to burn food chemicals and release energy in a process called respiration.

LATENT HEAT

Joseph Black's other research interest was the study of heat. He was curious as to why freezing snow could smother Edinburgh in a matter of hours, but took days to melt away. Black compared how long it took for a lump of ice to melt and warm up to the temperature of the room with the same weight of water that started at its freezing point but was already liquid. He discovered that the ice took 21 times longer. Black described this disparity as latent heat, or hidden heat, which was needed to turn ice into water rather than to make it warmer. It is now understood that latent heat breaks the bonds between the water molecules in ice, and as it does so there is no heat to spare to make the ice warmer – it stays at the freezing point.

Ice melts slowly, even in warm weather.

FLAMMABLE AIR

The next "air" to be found was discovered by Henry Cavendish, a young member of the Royal Society in London. He mixed iron filings with strong acids, and a very different kind of air bubbled out. While "fixed air" extinguished flames, the air collected by Cavendish burned very easily, often creating a squeaky pop as it did so. He noticed that droplets of water often formed on the inside of glass jars after he had set his discovery alight, but dismissed this as simply contamination. Cavendish announced his findings in 1766, and named the invisible, odourless and tasteless substance "flammable air".

At the time, the leading explanation about fire and heat was the phlogiston theory. This asserted that something burned as it released a diffuse, invisible, but still rather gooey and slow-moving fluid called phlogiston. This is explained why some materials lost weight after burning or heating – but did not explain

Henry Cavendish.

why others got heavier. Despite the flaws in the theory, phlogiston was a well understood concept in the 18th century, and Cavendish suggested his flammable air might actually be pure phlogiston.

PHLOGISTICATED AIR

Six years after the discovery of flammable air, another Scottish medical student found air that was the opposite of flammable. Daniel Rutherford was working on the principle that the air we breathed was a mixture of good and bad substances. The good air sustained life (and burning), and the bad stuff did not. He isolated a sample of air, and removed the good portion by introducing a mouse. He knew the mouse would die once there was not much good air left. Once that had happened, he burned a candle until it sputtered out, having removed the last of the good air. Rutherford bubbled the remaining air through lime water, which took out the fixed air. This he said was the source

Daniel Rutherford.

of the air's "badness", and removing it would make the air "good" again. Instead he found a lifeless air that seemingly did nothing at all. He called it "phlogisticated air" to indicate his theory that this air was saturated with phlogiston and was thus rendered inert.

1778 Carl Wilhelm Scheele discovers molybdenum, a name derived from the Greek word for lead.

1778 George Louis Leclerc, Comte de Buffon, estimates the age of Earth by timing how long a red-hot ball of iron takes to cool down. His theory is that Earth was once a hot ball and so extrapolates the cooling to account for a body the size of Earth, to arrive at an age for Earth of at least 75,000 years old.

1779 Jan Ingenhousz shows that photosynthesis does the opposite to respiration. It uses carbon dioxide and gives out oxygen.

1780 Antoine Lavoisier and Pierre-Simon Laplace build a calorimeter, a device for measuring how much heat is inside a material.

PHOTOSYNTHESIS

Antoine Lavoisier had the apparatus to show that the act of breathing in and out followed the same pattern as burning. Compared to what goes in, there is less oxygen more carbon dioxide coming out. When Jan Ingenhousz, a Dutch doctor, came to stay with Joseph Priestley at Bowood House, he discovered another life process that is just as important, and the total reverse of breathing. He revealed that plants take in carbon dioxide and give out oxygen. Today this is understood to be part of the process of photosynthesis, where a plant uses the energy in sunlight to turn carbon dioxide and water into the sugar fuels it needs to survive. Oxygen is a waste product, and without photosynthesis there would be no pure oxygen in Earth's atmosphere for us (and all the other animals) to breathe.

Sunshine is needed for photosynthesis.

DEPHLOGISTICATED AIR

Around the same time, a struggling preacher called Joseph Priestley was having more luck with chemistry experiments than in gathering a congregation. He succeeded in dissolving fixed air in water, and discovered that it made a very refreshing sparkling drink. He named it soda water, and parlayed this invention into membership of the Royal Society and an introduction to the wealthiest families in Britain at the time. (Johan Schweppes, a German jeweller, had another ideas. He took Priestley's soda water discovery and turned it into a mass-market beverage product that carries his name to this day!)

Priestley got a job as secretary to the Earl of Shelburne, and was given a laboratory for more experiments at Bowood House, the earl's country estate. He investigated the air given off by spirit of nitre (known today as nitric acid). When this invisible nitrous air was mixed in with air it spontaneously formed a thick red-brown fog, and the volume of gas reduced by a fifth. Priestley said that this was the "good air" – or dephlogisticated

Joseph Priestley.

air in the parlance of the day – which was able to accept phlogiston and thus support burning (and life). However, Priestley was unable to make a pure sample of this dephlogisticated air. In the end he found it by accident by heating red mercury. This mineral, actually orange, decomposed into pure mercury when heated, and Priestley found that as it did so it gave off a pure sample of his dephlogisticated air. This stuff made flames burn brighter and revived comatose creatures left to suffocate in bell jars.

TO FRANCE

Once Priestley had made his discovery public, he was summoned to Paris to meet Antoine Lavoisier. Lavoisier was an enormously wealthy French aristocrat, who spent his fortune on the biggest and best laboratory in the world, and was the leading chemist of his generation. His previous experiment had been to use vast lenses made from clear vinegar held inside a curved glass case to focus the sun's hot beams on a diamond, a pure crystalline form of carbon. The diamond got so hot that it was incinerated and became fixed air. Lavoisier felt sure it had combined

with Priestley's dephlogisticated air. Nothing could burn without that.

Lavoisier was able to show that dephlogisticated air and flammable air combined to make water. Water was not an element at all and neither was air. Lavoisier correctly surmised that these two "airs" were entirely new substances. He drew up an updated list of elements, which he called "simple substances". The list included carbon, sulphur, mercury and other metals known since antiquity. He added the new airs, naming them oxygen (meaning acid former) and hydrogen (meaning water

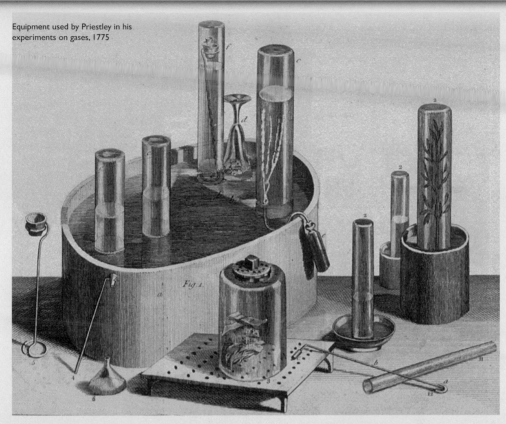

Equipment used by Priestley in his experiments on gases, 1775

former). The name for oxygen was based on Lavoisier's erroneous assumption that oxygen was the element at work in acids – it is actually hydrogen – but the name has stuck. He named phlogisticated air azote, which means "lifeless," but others preferred nitrogen, meaning "nitre former." (Nitre is another name for saltpetre, an ingredient in gunpowder.)

Lavoisier's list of elements was steadily built on, and within 25 years the study of mixtures of airs – by then known as gases – would be able to prove the existence of atoms.

Antoine Lavoisier.

Until the 1780s, electricity was largely a static phenomenon. Its effects were seen as an invisible force field that built up around charged objects and which attracted and repulsed in equal measure. It was possible to move charge as Stephen Gray had illustrated, and the Leyden jar was the best way of storing it.

JUMPING MONKS

In 1746, Jean-Antoine Nollet, a French priest and physicist, had shown both these concepts at work in one of the most mischievous scientific experiments in history. Using his influence as a clergyman, he dragooned 200 monks, arranging them into a human chain that formed a circle reportedly 1 mile (1.6 km) long. Each monk was connected to his neighbour through a short metal wire. Whether they knew what was going to happen next is unclear. Nollet connected the first monk to a charged Leyden jar and watched as the monks leapt in surprise as they all received an electric shock. Nollet had wondered whether the yelps of pain would proceed along the chain at a measurable rate, but in the end discovered that the electricity moved through the line of holy men so fast they all appeared to

Jean-Antoine Nollet.

get a shock at the same time. Electricity moved very fast indeed, it would appear.

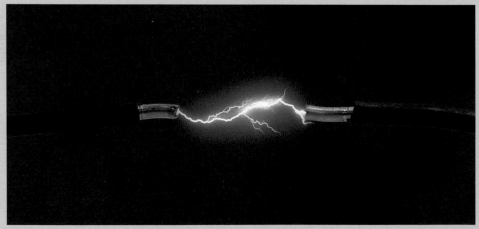

Spark caused by electricity between two wires.

1780 Luigi Galvani discovers that electricity is involved in muscle movements – and still works on dead animals!

1781 Tungsten metal, meaning "heavy stone," is discovered by Torbern Bergman.

• William Herschel, a German astronomer working in England, finds a sixth planet in orbit further away from Saturn. It is eventually named Uranus.

1783 The first flying machines to carry a human crew are two kinds of balloon. The first is a hot-air balloon and the second design is filled with hydrogen gas.

1784 The French astronomer Charles Messier publishes a catalogue of astronomical objects that do not appear to be stars. They include bodies now understood to be nebulae, globular clusters and galaxies.

• A teenage astronomer from northern England, John Goodricke, describes how Algol, a variable star, changes in brightness as a dimmer companion star orbits in front of it. He also discovers the first Cepheid variable, a kind of star that can be used to calculate interstellar distances.

Up, Up and Away

It is seldom noted that there were flying machines before we had invented effective cars. The first human aircrew took off in 1783 aboard a hot-air balloon made by the Montgolfier brothers, a pair of French paper merchants. Their 23-m (75-ft) tall bag made of varnished taffeta rose because it was filled with hot air, which took up more space than the same amount of cold air. Therefore the balloon floated upward like a cork in water. A few months later, fellow Frenchman Jacques Alexandre César Charles flew in a balloon filled with hydrogen (newly named). This gas is the least dense substance known; 13 times less heavy than air. Charles is also remembered for a second gas law, alongside Boyle's first one. Charles's law says that the volume of a gas is proportional to its temperature – something that the Montgolfier balloon relied on for its buoyancy.

Montgolfier balloon, the first practical hot-air balloon.

ELECTROSCOPE

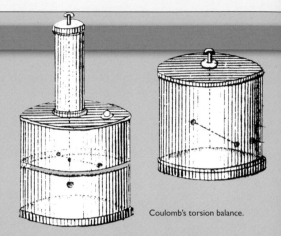

Coulomb's torsion balance.

Acouple of years later, Nollet built an electroscope, a device for detecting electricity. His was not the first. It worked like earlier versions. A metallic pointer moved toward or away from a source of charge. None were very reliable for comparing the magnitude of electrical effects. In other words, measuring electricity was proving difficult. In 1777, the French scientist Charles Augustin de Coulomb built a more precise device called a torsion balance, where a bar was free to move like the needle of a compass. The bar's motion was opposed by a twisted spring, and so large forces resulted in small movements. This made it possible to show up any differences in the magnitudes of one force over another.

By 1785, Coulomb had been able to use his device to discern a relationship between electrical forces and charge, now known as Coulomb's law. This says that a force from an object is proportional to the electrical charge on it and is inversely proportional to the square of the distance between charged objects. Physicists now measure electrical charge in units called coulombs. (One coulomb – 1 C – is the charge of about 6.2 million trillion electrons.) The coulomb is a lesser known electrical unit, because it does not measure charge in motion, or in a current. The electricity that we rely on to power our modern world is in constant motion which is measured in different ways.

ANIMAL ELECTRICITY

The first evidence of an electric current came from an unlikely place. In 1780, Luigi Galvani, a doctor and anatomy professor at the University of Bologna, made a chance discovery while dissecting the muscles and nerves in a frog's legs. When he hung up the freshly cut legs to drain using brass hooks on a cast iron railing, he was amazed to see the legs twitch as if brought back to life! Galvani recreated this set-up using a metallic arc tipped at one end with iron and copper at the other. Touching either end of a leg with this arc made the muscles contract. Galvani had made the first electric circuit, where an electrical charge flowed through the frog tissue and back around the metal connector.

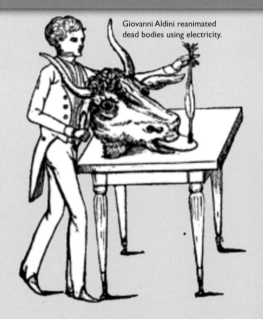

Giovanni Aldini reanimated dead bodies using electricity.

1785 Charles Augustin de Coulomb outlines the relationship between electrical force and electrical charge. He also shows that it is also inversely proportional to the square of the distance between charged objects. This relationship is known as Coulomb's law.

1787 The metal strontium is identified by William Cruickshank, and named after the nearby Scottish village of Strontian.

1789 Antoine Lavoisier draws up a table of the 33 known elements, or "simple substances".
• Martin Heinrich Klaproth discovers the heavy metal uranium and names it after the planet Uranus. He also discovers zirconium in the same year.

1791 Titanium is identified by William Gregor although the metal is not purified until 1910.

1794 Yttrium is discovered by Johan Gadolin, and named for the Swedish mining town of Ytterby. (This town eventually has three more elements named after it – erbium, terbium, and ytterbium.)
• Erasmus Darwin, the grandfather of Charles, publishes an early theory of evolution in a book called *Zoonomia*, which suggests that life has been changing since long before that dawn of human history.

1795 Joseph Bramah invents the hydraulic press which transmits and multiplies forces by pushing on incompressible liquids, such as oil or water. This forms the basis of hydraulic machinery.

Galvani declared that this "animal electricity" was the product of the vital force that permeated all living things, and thus produced in a different way to the charge generated by Hauksbee devices. The link between life and electricity, especially in the nerves and muscles, is indeed a crucial one, and Galvani's nephew Giovanni Aldini made that plain as he toured Europe putting on macabre shows where he reanimated corpses, even the bodies of recently executed convicts, using electricity delivered from Leyden jars. It is said that tales of these ghoulish exhibitions are what inspired Mary Shelley to write her story about Frankenstein's monster in 1818 – a monster that was brought to life with electricity.

Galvani demonstrates animal electricity with a frog.

CHEMICAL ELECTRICITY

Alessandro Volta.

However, the idea that animal electricity was somehow a special case unrelated to the electrical phenomena already known about did not sit well with some, especially another Italian scientist called Alessandro Volta. Volta correctly theorized that the key parts of Galvani's primitive electrical circuit were the metals, not the frog's legs. In 1799, he built an entirely lifeless version by piling up alternative discs of zinc and copper and separating each pair with a swab of wood pulp soaked in salty water. The metal discs took the place of Galavani's arc connector, while the frog's legs, juicy and fresh, were replaced by the sodden salty pulp.

When Volta connected the top of his pile to the bottom he was able to produce electric sparks and to demonstrate powerful electrical forces. The device became known as the voltaic pile and was the first example of the electric battery. (The term "battery" actually predated it, coined by Benjamin Franklin, who had likened a set of Leyden jars that had been assembled to supply large charge to a battery of cannons.)

POTENTIAL DIFFERENCE

So what was happening in Galvani's frog's legs and Volta's pile? In terms understood in their day, electrical fluid was moving from the copper to the other metal (iron or zinc were both used). The fluid was mediated by the salty liquid in the animal tissue and in the pulp. That movement of electricity created an excess of charge in one metal and a lack of it in the other. Charge always moves from a region of excess to one with a dearth of charge in order to even things out again, and so the difference in charge sees the "fluid" move from the zinc (or iron) to the copper again. However, that does not even out the charge, because the relative chemical properties of each metal means that zinc pulls on charge more strongly than copper. Therefore there is a persistent "potential difference" between the two metals which means that charge will always flow between them – or at

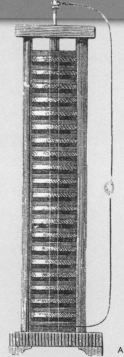

A voltaic pile.

1796 French anatomist Georges Cuvier's study of fossilized skeletons shows that ancient animals were different to the species that are alive today, and which had become extinct.

1798 Louis Nicolas Vauquelin identifies beryllium as a metallic element in emeralds and a related mineral called beryl.

1798 Henry Cavendish builds a giant torsion balance to measure Big G, the gravitational constant which relates the mass of an object to the force of gravity it produces.

1800 Alessandro Volta builds a chemical pile from zinc and copper. This is the first electric battery.

EXTINCTION

Fossils had been a source of dispute among biologists and naturalists. What did they represent? Some said they were tricks planted by God to test faith, others said that they were monsters destroyed in the early mythical phase of our history. Others wondered why God would make life just to destroy it, and they suggested that these were the ancient remains of animals that still live today. In 1796, the French anatomist Georges Cuvier showed that elephant-like fossils were not the same animal at all as the elephants that live today. Instead they were a different species, with a subtle but significant different shape to their skulls, skeleton and teeth. This was hard evidence that animals could become extinct and a species could die out completely. Cuvier believed that the extinctions were the result of the Great Flood as described in the Bible, and suggested there had been a wave of such catastrophes. Others took a broader view, and proposed that extinction was an indicator that life could change over time – or evolve.

Georges Cuvier.

least until the chemicals facilitating it run out. This concept of potential difference underwrites electric current. It is perhaps more familiar as the term "voltage", which is a measure of the force that is pushing the charge around a circuit – and is named after Volta. (Galvani is immortalized by the term "galvanize", which means to be shocked into action, as if by some inner spark.) Today we understand that the charge is moving as particles at the atomic and subatomic level, such as electrons, but it was another century before the idea of objects smaller than atoms could be considered. Indeed, the idea of atoms was still something under dispute – although that question would be settled very soon.

At the turn of the 19th century, the work of the pneumatic chemists had destroyed the last vestiges of classical science. Not one of the four elements – earth, water, air, and fire – which had formed the bedrock of thinking for more than two thousand years was, in fact, a so-called simple substance. Instead the list of elements was growing steadily, with vanadium, titanium, and uranium all newly discovered metals.

A new set of rules had developed to understand what an element was, a way of understanding that largely remains valid today: an element was a substance that cannot be separated into simpler constituents. Elements could combine to make a composite substance or compound. This was a completely distinct substance that frequently did not resemble its original ingredients at all. For example, hydrogen and oxygen, two highly flammable and invisible gases, had been shown to combine to form water, a liquid that cannot burn at all.

An atom is a collection of smaller, subatomic particles.

1800 The Industrial Revolution has begun in the United Kingdom, where automation, assembly lines, and steam-engine power are making mass production possible. This technology is also spreading across Europe and North America.

• William Herschel finds that when sunlight is split into its full spectrum by a prism, the hottest part is just beyond the red light at the edge of the spectrum. He names this invisible light infrared.

1801 The Spanish-Mexican naturalist Andrés Manuel del Río discovers vanadium compounds.

• While searching for a planet between Mars and Jupiter, Giuseppe Piazzi discovers Ceres, the first object seen in the Asteroid Belt.

• Thomas Young proves that light is a wave by making a distinctive interference pattern by diffracted two beams of light through narrow slits located side by side.

• Richard Trevithick, a Cornish engineer, builds a steam-powered road vehicle which carries passengers up the steep hill in the village of Camborne.

1803 John Dalton revives the ancient theory of atomism by suggesting elements are made from distinct units called atoms each with a specific weight. He also proposes that different substances are built from atoms bonded together in specific patterns.

• Palladium, the fourth and final precious metal (along with gold, silver and platinum) is isolated from samples of platinum by William Hyde Wollaston.

• The metal cerium is named after the recently discovered asteroid Ceres by the element's discoverer Martin Heinrich Klaproth.

ASTEROIDS

After the discovery of the sixth planet Uranus in 1781 by William Herschel, some astronomers compared the distances of the planets and found what appeared to be a big gap between Mars and Jupiter. A team of observers styled as the Celestial Police split up the sky between them to scour the background of stars for a new planet moving in the gap. One of them, Guiseppe Piazzi, spotted something on New Year's Day 1801. He watched it for 24 days, and its path suggested it was a planet not a comet. Then he lost sight of it, and the object, which he called Ceres, went missing. The Celestial Police asked for help from Carl Gauss, a young German who was soon to be lauded as the greatest mathematician of his generation. He took Piazzi's scant data and mapped out where Ceres would be once it circled the Sun. The astronomers finally found Ceres again on the last day of 1801. Further observation revealed that Ceres was too small to be a planet (it is about the width of Spain) and so it was termed an asteroid. Soon more asteroids were being found in the gap between Mars and Jupiter. Today the region is called the Asteroid Belt and has 50,000 recorded space rocks, of which Ceres is by far the largest.

CONSERVATION OF MASS

Many alchemists had based their thinking on transmutation, which was a way that elements could change from one form to another. Generally one or two elements were seen as dominant, and of course alchemists were most interested in getting other elements to transform into gold. However, another one of Lavoisier's contributions to chemistry was the law of conservation of mass. (He was dead by now, guillotined in 1794 by France's new revolutionary government for his close links to the hated *Ancien Régime*.)

The conservation law simply states that when elements react, the amount of matter before is identical to the amount of matter after the reaction. So the mass

Antoine Lavoisier.

INDIVIDUAL POINTS

The idea of atoms had been proposed by Leucippus and Democritus way back in the 5th century BCE. They arrived at the idea from a philosophical approach, and offered no actual evidence for it. That remained the case for the centuries to follow, although many scientists, from Boyle to Newton to Lavoisier, deployed the concept of atoms – also called corpuscles – to help illustrate their thinking on the nature of matter, heat and other phenomena.

In the 1730s, Daniel Bernoulli revisited the idea of atoms from a mathematical perspective, and devised a purely hypothetical system that explained the behaviour of air in terms of it being a collection of individual points of matter, each infinitesimally small, and each in constant motion following its own trajectory. This model of gas helped to visualize how invisible, diffuse matter could subject pressure on a solid surface, such as the inside of whatever vessel it was

Daniel Bernoulli.

stored in. (It would also help to understand the way a gas's temperature is proportional to the gas's pressure; hot gases exert higher pressures than cold ones. This relationship concerns the third gas law, Gay-Lussac's Law, but the reason for the link would not be clarified until the 1840s.)

of oxygen and hydrogen always equals the mass of water they make once they react. The same is true when the reaction is run backwards. (In the case of water, it can be split into hydrogen and oxygen again, but it takes a lot of energy.) During the reaction, the elements appear always to maintain their integrity. Lavoisier, like many chemists before him, had his suspicions about how that might be possible. In 1803, John Dalton, a teacher with a fascination for weather, was able to confirm it. While this discovery placed the last nail in the coffin of the classical elements, it resurrected another ancient theory, the idea that nature is made from atoms.

1804 William Hyde Wollaston isolates rhodium.
• The weaver Joseph Marie Jacquard invents a loom that can weave fabric using a pattern coded by a strip of punched cards. Punched cards become the earliest format for inputting data and programs to primitive computers.
• The American industrial engineer Oliver Evans builds the Oruktor Amphibolos, a steam-powered amphibious vehicle that drives on the roads and converts into a river boat. The vehicle's wheels fall off on its first journey though Philadelphia and it never reaches the water.

SOFTWARE

In 1804, a French weaver called Joseph Marie Jacquard invented a loom that could produce cloth with an exquisite pattern – all automatically and much faster than by hand. The pattern was produced by introducing particular threads to the weave in a particular order. That order was coded into a long string of cards with holes punched out of them in a specific pattern. When the cards were fed through the loom, each hole corresponded to the required thread. The loom was not a computer, but the punched cards were the first form of software. Swapping them with a new set meant that the loom produced a new cloth pattern. Punched cards were used to input data and programs to early mechanical computers and early digital devices in the the 1950s and 1960s.

Jacquard loom.

DALTON'S LAW

John Dalton, a self-taught scientist who spent most of his life living and working in Salford, a town close to the English city of Manchester, was able to draw together these different ways of understanding matter. In 1803 he set out his thinking with the first draft of the modern atomic theory. His journey to it had started in the 1780s, however, when as a young man he began to keep painstaking records of the meteorological conditions – the weather. (Meteorology is not exactly the study of meteors, but bears this name because ancient Greeks believed that shooting stars were atmospheric phenomena like thunder and lightning.) Dalton's daily measurements included the air pressure, which had been an indicator of weather changes since the days of Torricelli. While trying to forecast the weather, Dalton became sidetracked by the fundamental behaviour of the air.

By now he knew that air was a mixture of nitrogen and oxygen. (There was 1 per cent of the air that consistently defied definition, but was largely ignored as an error. It was later revealed to be argon – meaning the idle gas – and one of the inert noble gases along with helium and neon.) He was also aware of the gas laws that described the relationship between the pressure, volume and temperature of a gas. These laws worked in just the same way for air, a mixture of gases, as they did for a pure

John Dalton.

sample of one gas alone. Logically therefore, Dalton reasoned, the total pressure of air was made up of "partial pressures" exerted by the different gases. This idea is remembered today as Dalton's law.

Following on from this relationship among mixed gases, Dalton was able to explain that each gas acted independently of the others. When two gases were added to a container in equal amounts, they both spread out to fill the container evenly. At every point on the surface of the container, half of the pressure was exerted by one gas and the other half by the other. This was true no matter what the proportions of the gas mixture.

1807 Humphry Davy purifies potassium and sodium using electrolysis, a means of splitting up compounds using a powerful electric current.

1808 The French chemist Joseph Louis Gay-Lussac observes that gases combine in exact proportions and he finds that two parts of hydrogen react with every one part of oxygen. In the 1820s this ratio is expressed as the formula H_2O. Gay-Lussac also predicts the existence of boron, which is isolated nine days later by Humphry Davy. Davy also discovers calcium and barium.

DAVY AND ELECTROLYSIS

The voltaic pile offered scientists a new tool. It was already known that a powerful charge from a Leyden jar could be used to drive chemical reactions in reverse – a process dubbed electrolysis,

Sir Humphry Davy.

"splitting with electricity". In 1807, a young English scientist built the largest pile, or battery of piles, yet seen in the basement of London's newest science centre, the Royal Institution. Davy then used this powerful current to investigate common substances such as caustic potash and soda. These two "earths" were included on Lavoisier's and Dalton's lists of elements. However, Davy found that they both appeared to transform in the current into liquid metals that promptly caught fire. Potash produced a metal that burned with a purple flame, while caustic soda created orange flames. Davy had discovered the highly reactive metals potassium and sodium, and he went on to isolate magnesium, calcium, boron, and barium, making him one of the most prolific discoverers of elements in history.

PROPORTIONS AND COMBINATIONS

Dalton then sought to extend this physical behaviour of gases with chemical ones as well. He began to make known quantities of elements react together, such as hydrogen and oxygen, and carbon (as charcoal) and oxygen, and discovered that they always combined in fixed proportions.

The final piece of evidence considered by Dalton was that gases had distinct weights. He would of course have known that hydrogen was much lighter than oxygen or nitrogen, but his discovery about fixed proportions allowed him to calibrate more precisely the relative weights of each element.

A mural in Manchester's Town Hall of John Dalton collecting marsh fire gas.

Considering all this evidence – the partial pressures, even mixing, definite proportions when reacting, and the relative weights of gases and other elements – led Dalton to conclude that all elements were composed of invisibly small particles, the very same atoms that had been proposed by the ancient Greeks. The atoms of one element were identical to each other but different from those of other elements.

Dalton imagined that during reactions the atoms of different elements became bonded together, forming a cluster with a specific shape. There was already a word for something like this – molecule – which originally meant a tiny quantity of a substance but came to mean the smallest possible structure of a compound. Dalton's law of proportions told him that each element made a particular number of links with other atoms. He even built little wooden balls that could be connected with sticks to help him figure out how molecules formed. Some of his information about specific elements was not correct, but this general view of atoms and molecules was very powerful. It led to the next very big question: what is it about one element's atoms that makes it different to another's?

John Dalton's wooden atoms connected into molecules.

STUDYING LIGHT

At the start of the 19th century, there were two leading theories about what light was made of. The Islamic scientist Ibn al-Haytham had set the scene for the science of optics at the start of the 11th century. He took a geometric view which understood light as a series of beams moving in straight lines spreading out from the Sun, candle or other light source. The eye created vision from any beams that entered it, either directly or by first reflecting from an object. Al-Haytham's concept explains how light reflects off surfaces. The incident ray (from the Sun or other light) arrives at a specific angle to the surface. This angle is measured from an imaginary line, the normal, which is perpendicular to the surface. After the light hits the surface, it forms a reflected beam that moves away at the same angle as the incident one, only on the other side of the normal.

BENDING LIGHT

Other optical phenomena such as refraction could be described in a similar away. In refraction, a beam of light is deflected slightly as it moves from one transparent medium (air) to another (water). This is behind the common optical phenomenon where objects viewed through both media appear in slightly different positions to where they actually are. René Descartes, the French philosopher, mathematician and all-round genius who worked in the early 17th century, correctly surmised that refraction was due to differences in the speed at

Optical refraction.

which light moved through each medium. (Light moves fastest through a vacuum, and famously nothing can go faster than that. However, it does slow down a little in air, glass and water etc., and each one of these media has a specific speed of light.)

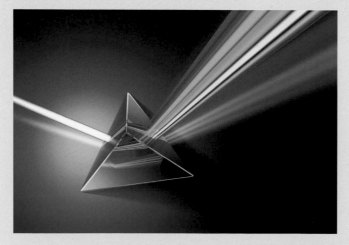

White light dispersions,

LAMARCKISM

Fifty years before Darwin's theory of evolution by natural selection, the Frenchman Jean Baptiste Lamarck had already proposed a theory of how animals and plants could change over time. Lamarckism says that organisms changed because the characteristics they acquired in life were passed on to their offspring. The classic example is the giraffe, which evolved due to many generations of giraffes stretching to get at the leaves beyond the reach of other animals. That stretching makes the neck grow longer, and the giraffe babies are born with longer necks. Darwin's simple but grim tweak was that some giraffes simply have longer necks than others – and the short-necked ones starve to death.

19th-century picture of a giraffe.

1809 Jean-Baptiste Lamarck proposes a theory of evolution in which animals change by inheriting characteristics acquired during the life of their parents.

1810 After Frenchman Nicolas Appert develops the means of preserving foods inside sealed glass jars, Englishman Bryan Dorkin opts to seal food inside iron cans, and sets up the world's first commercial cannery.

1811 Iodine found in the ash of seaweed by Bernard Courtois.

1812 Amedeo Avogrado concludes that a fixed volume of any gas contains the same number of atoms or molecules.

AVOGADRO'S NUMBER

In 1811, Amedeo Avogadro said that since gas pressure, temperature and volume are always linked, that means that two different gases with the same pressure, volume and temperature must therefore contain the same number of particles. It took another 50 years to realize the importance of it, but this idea was used to create the periodic table of elements, one of the most powerful classification systems ever devised.

CORPUSCLE THEORY

In the 1660s, Isaac Newton postulated that light beams were streams of tiny solid particles – or corpuscles, as was the term of the time. He was not the first to have this idea, but he added weight to the theory by proposing that the corpuscles bounced around fully in accordance with his laws of motion. True to form, Newton did not really tell anyone about his idea for the best part of 40 years, and

WAVES IT IS

Newton's idea worked well for reflection but less well for refraction. Meanwhile, Huygens's theory fitted perfectly. The particle theory of light really fell down when to came to diffraction. Light diffracts when it shines through a very narrow slit of tiny hole. The light beams that make it though do not maintain their straight path, but spread out as they pass through the constriction, effectively making two new point sources of light. Waves of all kinds – sound, water ripples, etc. – do this. Streams of particles do not.

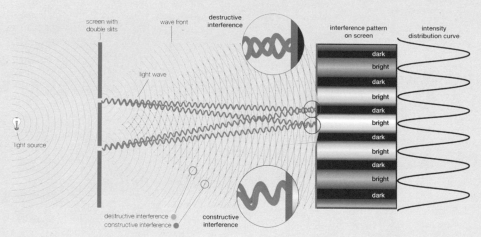

Interference pattern of Young's double slit experiment.

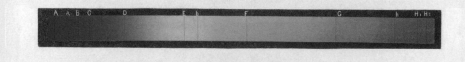

in 1670s, the Dutch scientist Christiaan Huygens proposed that light was in fact a wave that propagated outward in all directions.

Nevertheless, the corpuscle-versus-wave debate continued for another century, often lining up on nationalistic grounds. However, in 1801, the English researcher Thomas Young performed an experiment that proved that light was a wave. He shone light through two slits located side by side. The diffracted beams were then projected on to a screen and they made a distinctive striped pattern of light and dark regions. Such a pattern could only be caused by interference, which is a wave-only behaviour. When two waves meet they interfere by merging together. If they are in perfect synchrony, rising and falling at the same time, the two waves merge into a single larger wave, making a brighter light. If they are out of sync, the waves cancel out each other, making only darkness. Young's interference pattern was proof that light is a wave.

1813 Jöns Jakob Berzelius introduces formulae (e.g. H_2O) to denote chemical combinations.

1814 Joseph von Fraunhofer notices dark lines in the spectra coming from stars, founding the science of spectroscopy.
• George Stephenson builds the first practical full steam locomotive at the Killingworth Colliery in northeast England.

1815 Humphry Davy invents a safety lamp for miners. A naked flame in a mine could ignite explosive gases released from workings. The Davy lamp shrouded the flame in a copper gauze which prevented heat from reaching the air – and any other gases – outside.

1816 The first photographic negative is made by Nicéphore Niépce using a light-sensitive paper in a camera obscura.

1817 Cadmium is identified as a new element by Germans Friedrich Stromeyer and Karl Samuel Leberecht Hermann, each working independently.
• Lithium is discovered by Johan August Arfwedson.

1819 Scottish engineer John Macadam develops a fast and inexpensive road-building system that covers roads with a layer of crushed stones. This is known as macadam. In the early 1900s tar was added to the system to make tar macadam or tarmac.

MISSING COLOURS

In 1814, a German lens-maker called Joseph von Fraunhofer built a device that would lead to the next century of discoveries about light. He developed glass for lenses that reduced the blurs and aberrations found in earlier optical devices. His prisms were so precise that he could examine the rainbow of colours created through a magnifying glass. He saw that there were distinct dark lines or gaps among the blend of colours. These so-called "Fraunhofer lines" would later be used to discover elements, classify stars and reveal the internal structure of atoms.

Until the 1820s, the phenomena of magnetism and electricity, both known since ancient times, were thought to be entirely separate. There were similarities, of course, both were capable of producing attractive and repulsive forces, and there were hints at other links. For example, when a house was struck by lightning, the iron knives in the kitchen were frequently left magnetized.

CHANCE SWING

Nevertheless, magnets were of little interest to scientists. However, the new phenomena of electric charge, batteries and electric currents, did attract attention. In April 1820, a professor in Copenhagen, Denmark, called Hans Christian Ørsted was giving a lecture on the heating effect of electric currents in metal wires. While demonstrating this using a voltaic pile from his lab, a compass needle on the desk beside him swung to point at the electrified wire.

This showed Ørsted that the electrical current was making the wire magnetic, but only temporarily. When the current was switched off, the needle swung back to point north. Ørsted needed to discount the effects of heat, and repeated the

CATALYSTS

In 1823 the German chemist Johann Wolfgang Döbereiner produced a lamp that burned by hydrogen gas. The gas was made inside by reacting zinc with sulphuric acid, and then it passed through a platinum gauze. As if by magic, the gas bursted into a bright flame. Without the platinum, the hydrogen would simply escape – which is never a good idea. This curious phenomenon was given its name

Döbereiner's lamp, the first modern lighter.

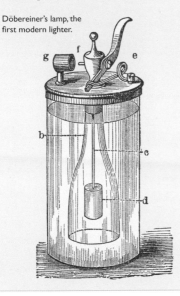

1820 Danish professor Hans Ørsted reveals a link between electric current and magnetic fields.

1821 André-Marie Ampère uses Ørsted's discovery to create a new field of physics called electromagnetism. He ascertains the behaviours of electromagnetic forces.
• The thermoelectric effect, where heated conductors produce electric current, is discovered by Thomas Johann Seebeck.
• Michael Faraday invents a homopolar electric motor which turns electricity into motion.

1822 The first fossil identified as a dinosaur is unearthed in England by Gideon Mantell.

1823 The first large-scale public water treatment plant opens in Chelsea, London.
• Johann Wolfgang Döbereiner builds a lamp and lighter that uses a platinum catalyst to burn a source of hydrogen gas. (The general concept of a catalyst is described by Berzelius in 1836.)

13 years later by the Swedish chemist Jöns Jacob Berzelius. He described it as catalysis from the Greek word for "untie". The platinum was a catalyst, which made hydrogen react with air without the need for a source of heat to start it off. We now know that a catalyst brings reacting substances close together on its surface, reducing the amount of energy they need to react.

process using thinner wires. These got hotter but their effects on the compass were more feeble. Completely by accident, Ørsted had discovered an inextricable link between electricity and magnetism, and as such founded a field of physics called electromagnetism.

Hans Christian Ørsted.

FORCE FIELDS

The French researcher André-Marie Ampère quickly developed a much clearer picture of electromagnetic forces. He showed that two electrified wires could repel and attract one another in the same way as a pair of magnets. The polarity of the wire depended on the direction of the current. Currents running in opposite directions through a pair of wires caused a repulsive magnetic force, while currents running in the same direction made the wires attracted to each other.

It is a testament to Ampère's work that the unit of electric current, the amp (short for ampere; A) is named for him, not any of the many other figures who investigated the phenomenon. Another of those figures was the English scientist Humphry Davy, who had become the world's first scientific celebrity due to his entertaining public lectures and prolific tally of new elements that he'd discovered

André-Marie Ampère.

using electrolysis 15 years before. Davy's chief collaborator was William Wollaston, and it was he who had overseen the construction of the huge battery beneath the Royal Institution which Davy used. Another name thrown into the mix was Michael Faraday. One day soon the name Faraday would eclipse all others in the field of electromagnetism, but to begin with he was a self-taught amateur scientist who worked as Davy's laboratory assistant.

MOTOR QUEST

Soon after hearing of Ørsted's discovery and then Ampère's account of electromagnetic forces, Davy and Wollaston began to ponder whether it would be possible to use the repulsive and attractive forces of electricity and magnetism to create rotational motion. In 1821, Faraday, their trusted underling, beat them to it, and he is credited with the invention of the world's first electric motor, albeit a very rudimentary one.

Faraday's device was called the homopolar motor. He had used a compass to map the direction of the forces that surrounded an electric conductor, and that showed that the wire had a circular force field around it. His motor plan was to have the wire's force field push against the magnetic field of a static magnet and

Humphrey Davy.

create a continuous motion. To allow the wire to move, its tip was suspended in a small dish of mercury, which completed

1824 The Swedish chemist Jöns Jakob Berzelius isolates pure silicon from silica, the main mineral in sand. Silicon is the second most common element in rocks after oxygen.

• Nicolas Léonard Sadi Carnot, better known as Sadi, develops the Carnot Cycle which explains how a heat engine converts thermal energy, or heat, into kinetic energy, or motion.

• William Sturgeon invents the electromagnet and builds a functional electric motor.

1825 Bromine is discovered by Antoine Jérôme Balard. This is the only non-metal element that is a liquid at room temperature. Its name is derived from the greek word for "stench".

• Aluminum is purified by Hans Christian Ørsted using electrolysis.

HEAT ENGINES

While others were developing new electrical engines, the French scientist Sadi Carnot wanted to better understand the older kind of steam engines. He figured out that the thing that made them move was heat, carried by a fluid – steam in this case. The heat moved from the hot steam to the outside of the engine, and as it did so some of it was transformed into another more useful form – motion. To make the engine run faster and with more power, one needed to increase the difference in temperature between the heat source (the boiler) and the heat sink (the outside).

Industrial steam engine.

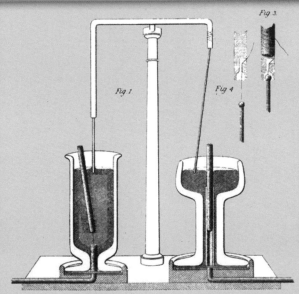

the circuit. When a magnet was placed in the centre of the dish, and the current turned on, the wire whipped around it in a circle. Although Faraday did not understand it in this way, the chemical reactions in the battery created a surge of electrical charge through the wire, and this surge was being converted into mechanical motion. Today's electric motors do pretty much the same thing, although one of the elements – generally the conductor – forms an axel or drive shaft that transmits the rotational energy out to a drive wheel or similar actuator.

Faraday's electric motor – the homopolar device.

ELECTROMAGNETS

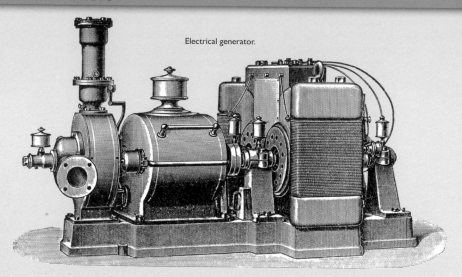

Electrical generator.

Davy never really forgave Faraday, and so the younger man was forced to keep his further research into electricity very quiet – until 1829, when the older one died. In 1824 William Sturgeon invented the electromagnet. This was a core of iron with a coil of copper wrapped around it. When a current ran through the wire, the iron core became a powerful magnet. When the current was switched off, the magnetism disappeared. The electromagnet became part of Faraday's research, and Sturgeon used it to refine a more practical electric motor design. The electromagnet offered something else – action at a distance. It was possible to control an electromagnet along a wire which could be miles long. This switching system formed the basis for the first telegraphs – and electromagnets are still used in radios, loudspeakers, microphones and even computer hard drives.

19th-century magneto electric machines.

INDUCING A CURRENT

However, electromagnets also featured in Faraday's biggest discovery of all – one that resonates to this day. He knew that a simple wire coiled around iron made an electromagnet, but what happened if there were two coils? To find out, he added two coils to an iron ring. When one coil carried a current, Faraday detected a tiny flicker of electricity in the other. Further investigation showed that the current in the second coil was being induced by a change in the magnetic field that was created by the appearance of current in the first. Faraday called this phenomenon electromagnetic induction. (He shares the discovery of induction with the American Joseph Henry who saw something similar around the same time. Henry's work on electromagnetic relays paved the way for electronic telecommunications, but that is another story.)

Faraday realized that induction could turn the motion of a magnet into a flow of current in a conductor – the precise opposite to the action of an electric motor. Today's electricity generators, which power the modern world, do just that. A conductor is placed within a powerful magnetic field of large electromagnets. It is then made to rotate by a turbine

Michael Faraday.

1827 The Scottish botanist Robert Brown observes apparently random movements of tiny pollen on the microscopic scale. Almost a century later, Albert Einstein explains that this "Brownian motion" is due to the vibrations of atoms.

BROWNIAN MOTION

Despite all the great works in the physical sciences at the time, the 1820s was the age of the life scientists. After all, this was a time when intrepid explorers were returning from all Robert Brown. *corners of the planet with mysterious creatures and miraculous plants. In 1827 a Scottish botanist called Robert Brown was studying the pollen of a plant from the Pacific Northwest, and found that the tiny grains jiggled randomly under the microscope. This phenomenon was recreated with all kinds of minute objects, from fungal spores to coal dust. It was recorded as Brownian motion, and 80 years later Albert Einstein explained what it was. It was the first visual proof that atoms exist. The grains are being jiggled by the random motion of atoms – atoms that can never be perfectly still.*

set spinning by wind, fast-flowing water, or steam. As the conductor moves, the orientation of the magnetic field is constantly changing and that induces a current in the conductor. Within a few seconds (it's moving about half the speed of light), that current becomes available at the plug sockets in homes.

IONIC BONDING

Even after transforming the world of physics and electrical technology, Faraday was not nearly over. His next contribution (but not his last) was to take a closer look at electrolysis, the analytical technique that had made the name of his former boss. He found that the quantity of material decomposed by the electricity was proportional to the size of the current. As with the battery and his motor, his idea was to do with energy. The electrical energy was being transferred to the compounds, causing them to split into simpler constituents.

As a supporter of the one-fluid theory of electricity, Faraday suggested that electricity was being carried through liquids by charged bodies, which he named ions (from the Greek for "mover"). Cations were positively charged and they were attracted to the negatively charge electrode, or cathode. The anode, or positive electrode, attracted the negatively charged anions. The molecules of compounds, Faraday suggested, were built from anions and cations that were bonded together by mutual electrical attraction. These terms, cited by Faraday and his close colleagues, remain in use today, and his theory of ionic bonding is the basis of some chemical structures. But not all. To understand more, chemists were going to have to deconstruct the atom itself – and that would have unforeseen effects.

ORGANIC CHEMISTRY

Until 1828, scientists thought that living things were powered by some kind of vital force which was different to the processes that were at work in the inanimate world of chemicals. It was impossible to make the materials found inside bodies in a laboratory – that required the action of the vital force. Then the German Friedrich Wöhler made urea by accident (he wanted to make something called ammonium cyanate). Urea is the primary constituent of urine. Wöhler's chance discovery was the first hint that the chemistry of living things followed the same rules as any other kind of material.

Friedrich Wöhler.

1828 Friedrich Wöhler accidently proves that chemicals used by living bodies can be made in the laboratory by non-living process. This shows that life uses regular chemistry not a separate "life-force".

1829 Jöns Jacob Berzelius discovers thorium, naming it after the Norse god of thunder, Thor.

1830 Jöns Jakob Berzelius suggests that atoms are held together by electromagnetic forces. In 1834, Faraday refines this idea by introducing the concept of ions.

• Jacob Perkins, an assistant of Oliver Evans, uses his mentor's original idea to build an ice machine in London. He takes out the first patent for a refrigeration system.

• The Liverpool and Manchester Railway, built by George Stephenson and his son Robert, becomes the first inter-city rail passenger service in the world.

• Charles Lyell publishes *Principles of Geology*, which publicizes the concept of uniformitarianism and highlights the great age of Earth among the scientific community at large.

1831 Michael Faraday discovers electromagnetic induction, where an electric current forms in a conductor when it moves through a magnetic field. He shares this discovery with American Joseph Henry. Faraday uses it to invent the electricity generator, while Henry develops the electromagnetic relay which is an integral part of the telegraph and later forms of telecommunication.

THE FRIDGE

In 1748, the Scottish chemist (and Joseph Black's teacher) William Cullen used a vacuum pump to make gas expand very fast. As it did so the temperature dropped, and the water around it froze. In 1805, the American engineer Oliver Evans proposed a cycle of expansion and compression in a closed system to create a constant supply of this cold. In 1830, his assistant Jacob Perkins put that plan into action, won the patent for it, and built the world's first refrigerator. It did not work very well, and took another 90 years to perfect, but has since altered the way food is transported and stored, often keeping items "fresh" for many months after they were taken from the field or slaughterhouse.

Jacob Perkins.

THE AGE OF EARTH

In the 1830s, after quietly accruing centuries of evidence from all quarters of the scientific community, Charles Lyell launched a new science, with the publication of *The Principles of Geology*. This book was a major influence on Charles Darwin and others who sought to understand where Earth's fascinating biodiversity came from. However, Lyell himself owed much to earlier investigators – perhaps going right back to the beginning.

WRITTEN EVIDENCE

Science began with "natural philosophers" such as Thales and Anaximander wondering how (and perhaps why) nature was in a constant state of change. Of primary interest was how it might change in the future – would the changing ever stop, for example, and if so, when? How long the process of change had been going on already was of lesser importance. That was a question dealt with by creation of myths and religion, and few felt the need or the courage to question that. Indeed, in 1654, James Ussher, the

A volcanic eruption.

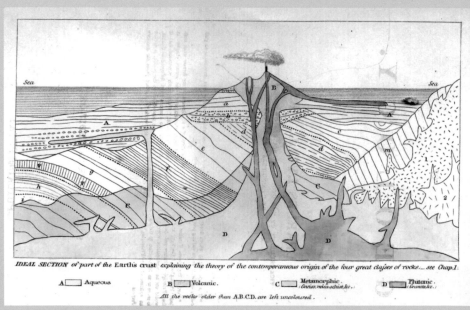

The frontispiece from Charles Lyell's *Principles of Geology.*

James Ussher.

1833 Anselme Payen discovers diastase in saliva. This is the first known enzyme, a biological catalyst. Specifically, diastase breaks up starch in foods into simpler sugars.
• Charles Lyell publishes the final volume of *Principles of Geology*, which popularizes the foundational ideas of earth science.

1834 Michael Faraday proposes that compounds are formed from charged particles held together by electromagnetic forces. He calls the particles ions.

1835 Artillery engineers had long reported that projectiles appeared to curve in flight. Gaspard-Gustave Coriolis explains that this is due to an apparent force caused by the rotation of Earth. Now called the Coriolis effect, it is also responsible for the deflection of winds and ocean currents in gyres.

1837 After the success of his Difference Engine mechanical calculator, Charles Babbage presents designs for an Analytical Engine, which is capable of being programmed and has an internal memory. As such, this device, although never built, is regarded as the first computer.

chief Catholic clergyman in Ireland, used the chronology contained in the Old Testament to count back the years to the literal moment of Creation as begun in Genesis, Chapter 1, Verse 1. Ussher's conclusion was that Earth was formed on October 22, 4004 BCE.

SET IN STONE

The following decade Niels Steensen, better known by the Latinized form Nicolas Steno, had noticed that the teeth of living sharks had an identical anatomy to "tongue stones," which had been collected since antiquity. Pliny the Elder, the Roman naturalist, had suggested these objects had fallen from the sky,

Fossil of an ancient crocodile.

Fossilized shark tooth.

but Steno contended they were shark's teeth that had been turned to stone – and all while buried in solid rock deep underground. Steno's clearly expressed and evidenced idea added weight to other voices – from Leonardo da Vinci to Robert Hooke – that fossils were the remains of once living organisms. The next question was, how long did it take to become a fossil?

THE AGE OF EARTH

COOLING DOWN

In 1778, Georges-Louis Leclerc, Comte de Buffon, who was the director of Paris's botanical gardens, devised a test to see how old the Earth might be. Isaac Newton had developed his own temperature scale, but the reason why we do not measure things in degrees Newton was that his system was devised to tackle hot objects, like coals and molten metal. The Newton scale was deeply flawed but it led to a better understanding of the rate at which hot objects cooled. (Newton could always be relied on for a mathematical breakthrough.) Buffon reasoned that Earth had started out as a red-hot ball of rock and metal and had been cooling ever since. So he scaled down the problem to a small sphere of solid iron, which he heated up and then timed how long it took to cool again. That time he then extrapolated to account for the very much greater size of Earth. His result was that Earth was 75,000 years old. To modern eyes, this is way off the true figure of 4,500 million years, but it nevertheless projected the past far beyond the dawn of recorded human history, and thus opened

Georges-Louis Leclerc, Comte de Buffon.

up the possibility that much of Earth's past did not involve humans at all. The Catholic Church ordered Buffon's books to be burned.

CROSSHEAD

However, in Scotland James Hutton was entertaining similar ideas about the great age of Earth. Nearly three decades of research resulted in him presenting his ideas to the Royal Society of Edinburgh in 1785 – sponsored by his friend Joseph Black, still something of a supremo in the Scottish Enlightenment. Hutton's big idea had a big name: uniformitarianism. Put simply, this is the concept that the processes we see happening today are the same ones that created the ancient formations of Earth over its long history.

So evidence for things like water and wind erosion, volcanoes and earthquakes, can be seen in the many layers of rock that make up Earth's surface. Right now, a new layer of rock is being formed from the soil and sediment that covers the surface, only it takes much longer than any human lifetime. A fellow Scot, Charles Lyell, extended and popularized Hutton's ideas in *Principles of Geology*, and it achieved extra resonance from a new line of research that had begun in England a few years before.

TERRIBLE FINDS

In 1822 the fossil hunter Gideon Mantell was working in a quarry near the south coast of England, when he dug up the skeleton of a mighty 10-m (33-ft), four-legged creature. His attention had been drawn to the area by a tooth fossil which seemed to belong to a lizard, only it was way too large to be found in the mouth of any species he knew. He named the creature iguanodon, which means "iguana-tooth".

Skulls and bones of similar giant reptiles had been known about for centuries. In China they were known as dragon bones. However, in 1841, the English biologist Richard Owen gave them a much better name: dinosaur. This comes by combining the Greek word for lizard, "sauros," with "deinos," which is the Greek for "terrible" or "fearsome".

1838 Carl Gustaf Mosander identifies lanthanum, a rare earth metal.
• Equipped with a telescope built by Joseph Fraunhofer, Friedrich Bessel is able to measure the parallax of stars (their apparent shift due to the motion of the observer) and so calculate the distance to the stars. He introduces the light-year unit to measure these great distances.
• The SS *Great Western* is the first steamship large enough to make a regular transatlantic passenger service.
1839 A Scottish blacksmith, Kirkpatrick MacMillan, invents an early form of peddle-powered bicycle.
• Jan Evangelista Purkinje discovers fibres in the heart which control its beat.

Iguanodon.

Since the Classical era, heat had been regarded as something substantial. Not quite one of the four elements, but an associated "virtue" or "principle" at least. Fire and air had more of heat than water and earth, yet it was debatable whether the chilly latter two elements merely lacked the virtue of heat or were instead filled with an opposing frigorific substance. Nevertheless, the idea of heat as a substance persisted well into the Scientific Revolution. Indeed,

Fire was originally regarded as one of the elements..

MEASURING HEAT

Lavoisier and his colleague Pierre-Simon Laplace invented a device for measuring the amount of caloric in a substance. The calorimeter, as it was named, had a central insulated chamber within which a substance was burned. Any heat released in the combustion melted a quantity of ice that was packed around the chamber. The volume of meltwater collected was an indication of the heat released. Modern calorimeters still use the same principle — although they heat water instead of melting ice — to measure the energy content of foods. (The unit calorie came a few decades later but was essentially derived from the work carried out by Lavoisier and Laplace in the 1780s.)

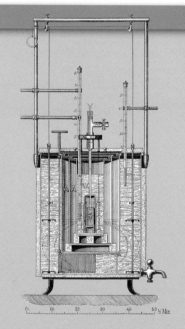

Calorimeter.

Antoine Lavoisier's list of "simple substances," which closed out the 18th century as the definitive description of natural materials, included heat — termed "caloric" — as one of the elements. (Light was also listed.)

BORING EXPERIMENT

In 1798 the American Benjamin Thompson - although now styled as Count Rumford, having skipped to Europe during the Revolutionary War - wanted in investigate why rubbing an object made it hot. This did not sit well with heat being an invisible fluid that oozed from one place to the next. While working as a weapons expert in Bavaria, Rumford performed a very significant experiment. He placed a cannonball in a barrel of water. He then bored into the cannonball with a drill - he selected the bluntest bit he could find. The friction of the drill bit on iron ball resulted in heat, and in two and a half hours of drilling the water around the ball began to boil. No matter how long he continued, the supply of heat was seemingly inexhaustible. This suggested to Rumford that heat was not a fluid, which would surely run out, but was instead linked in some way to the motion of the drill.

Rumford's boring experiment failed to excite the rest of the scientific community, and the subject of heat as motion did not reappear until the early 1840s. Around this time a young German doctor called Robert Mayer was working his passage to the East Indies aboard a Dutch ship. He noticed that the blood of the sailors he stitched up became brighter red as the voyage entered warm tropical waters. Bright red blood tells a doctor that it is full of oxygen. That could mean it comes from deep within the body, and this indicates a serious injury. However, Mayer's patients were bleeding

1840 Louis Agassiz proposes that Earth has experienced an ice age in the past.

1841 A huge volcano, named Mount Erebus after the ship of its discoverer Sir James Clark Ross, is found near the coast of Antarctica.

1842 Christian Doppler publishes the phenomenon now called the Doppler effect. It explains the way waves are compressed or stretched by the relative motion of their source and observer. The effect can be noticed in light and sound.

• The doctor Robert Mayer proposes the first law of thermodynamics which says that energy is always conserved, never created nor destroyed.

1843 James Prescott Joule calculates the mechanical equivalence of heat, or how motion energy is related to heat energy.

• The Brunels, Marc and Isambard Kingdom, a father and son team, dig the the Thames Tunnel under London's river. It is the first tunnel to pass through soft soil using a protective tunnelling shield.

Count Rumford.

bright blood from shallow cuts. Mayer realized that in the warm weather, their bodies needed to burn less fuel and use less oxygen to keep the body alive.

CONSERVATION OF ENERGY

Mayer began to think about the way the "living force" flowed through the body. It was contained in food, which was burned up to power the work of the body and given away as heat. The "living force", which by now was being described as "energy", was not created in the body but simply converted from one form to another – from chemical energy to motion energy and heat energy.

Mayer published this idea in 1842. It is now referred to as the conservation of energy and is the first law of

Julius Robert von Mayer, German physician.

MOTION EQUALS HEAT

The beauty of Joule's idea was that energy is not just about lifting and shifting mass around, but the same process is producing heat, electricity, sound, and chemical activity. It can also be stored inside material as potential energy. Joule was able to demonstrate this with an experiment to measure the "mechanical equivalent of heat". It was much like Count Rumford's but enabled Joule to measure how much heat was produced by motion. He fitted a rotating paddle inside a tank containing 0.45 kg (1 lb) of water, and attached the paddle to a weight. Every time the weight was dropped, the paddle spun around. The motion of the paddle added energy to the water, making it warm up very gradually. Joule discovered that the heat energy needed to warm the water by 1 degree Fahrenheit was the equivalent of moving a 0.45 kg (1 lb) weight 222 m (728 ft). The link between the motion of atoms with heat and temperature (and pressure) allowed scientists to rewrite our understanding of physical and chemical properties. That process has continued to this day via the work of Einstein to the hunt for the Higgs boson.

Joule's heat-equivalence experiment from the 1840s.

thermodynamics, the area of physics that describes the behaviour of heat and other energy. However, Mayer was universally ignored. And to his despair, the following year an English scientist called James Joule presented much the same idea, and did so in a way that was embraced by the scientific community. Today we measure energy using the joule unit (J) not the mayer. (One joule is the energy used to accelerate one kilogram to a speed of one metre per second in a second.)

1844 Samuel Morse sends the first telegram from the Capitol of Washington, DC, to a railroad depot in Baltimore. Using Morse's code of dots, dashes and spaces, the message spells, "What hath God wrought."
• The Swedish chemist Gustaf Erick Pasch invents the safety match, which uses sulphur in the head and red phosphorus on a rough striking surface.

1845 The Irish aristocrat William Parsons builds the largest telescope of the 19th century in the grounds of his estate. The telescope is nicknamed the Leviathan and it is used to observe the first spiral galaxy.

1846 Neptune is discovered following predictions of its orbit made by mathematician Urbain Le Verrier.

NEPTUNE BY NUMBERS

After more than 50 years of observation, it was becoming clear that the orbital path of Uranus did not match with the one calculated by astronomers. The wobbles in its motion indicated the presence of another large body further out from the Sun. Galileo had actually seen this body but not recognized it for what it was: the eighth planet. In the end the planet was not located by observation but by mathematics. The Frenchman Urbain Le Verrier calculated where the planet would be from its effect on Uranus, and in 1846 his predictions bore fruit. The new planet glowed with a faint ocean blue, and was named Neptune after the god of the sea.

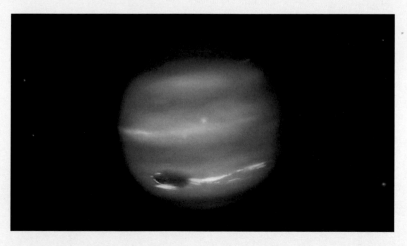

Biology was the ignored member of the scientific family for much of the Enlightenment. While chemists and physicists were debunking ancient beliefs and steadily revealing the hidden mysteries of the physical world, the old myths about biology clung on. For example, Jan Baptist van Helmont, an early researcher into gases and one of the pioneers in the Scientific Revolution, recorded in his writings a recipe for mice. He instructs us to put some old rags in a pot and sprinkle in some grains. Leave for 21 days and then you will find adult mice, including already pregnant females. His suggestion was that the mice have spontaneously appeared!

A caricature of Charles Darwin, following publication of his *On the Origin of Species*.

1847 James Young Simpson starts using chloroform as an anaesthetic during childbirth.

1848 Lord Kelvin, then William Thomson, proposes the Kelvin temperature scale which begins at 0 K, or absolute zero. At absolute zero, the heat energy on a substance is at its minimum level. It is not possible for a substance ever to get this cold – but it can get very close.
• Phineas Gage, a US railway worker, survives after a 1-m (3-ft) long spike is blasted through his head. This case is a seminal event in the study of the functional anatomy of the brain – which bits do what. It also reveals that the organ is highly plastic, meaning it can alter its function and organization throughout life.

1849 The first accurate measure of the speed of light is made by Hippolyte Fizeau. He achieves this by reflecting a light beam through a spinning cog and matching its rotation with the flickering of the light.
• James Francis invents the water turbine which is now the basis of hydroelectric power.

1851 Léon Foucault uses a pendulum to provide proof that Earth is actually rotating.
• Heinrich Schwabe suggests that sunspots appear according to an 11-year cycle.

1852 The Bunsen burner is invented by Robert Bunsen for burning samples in a clean flame.

1853 The hypodermic syringe is invented independently by both Charles Pravaz and Alexander Wood.

1854 Dr John Snow maps the incidence of a cholera epidemic in London's central Soho district. He finds that all households afflicted by the disease collect their water from a pump on Broad Street, and this is source of the illness. This marks the start of the science of epidemiology, the study of the spread of disease.

1855 Henry Bessemer develops a new batch process for producing high-quality steel in large quantities.

FOUCAULT'S PENDULUM

In 1851, Léon Foucault set up a long pendulum in Paris. As it swung it dragged the pointed tip of the bob through a layer of fine sand on the floor. At first the pendulum appeared to swing back and forth, but over the following hours it appeared to rotate clockwise. Pendulum law says that the motion of the bob is always in the same direction. So what was going on? After 24 hours, the pendulum had returned to its original plane. It was not the pendulum that had turned but planet Earth beneath it. Foucault's pendulum offered the first direct evidence of Earth's rotation. It turned out that Copernicus, Kepler and all those other astronomers were right all along.

Foucault's pendulum in Paris.

HIERARCHICAL VIEW

This idea of spontaneous generation is an ancient one, probably emerging from our most primitive ideas about life. Aristotle put it down to a "vital heat" that was present in all material, which coalesced into life forms. First off, small critters, like worms and insects, formed directly from inanimate soil, and then these steadily developed in complexity over many generations until they formed into the big animals - like us. This was in keeping with Aristotle's teleological approach to nature. He believed that natural processes were being drawn towards a final state of perfection. That included the development of life, which formed a Great Chain of Being. As one might expect, humans were near the top, jostling for position with the gods.

Such thinking persisted through the Middle Ages. Barnacle geese from the north and west of Europe were so named because they were believed to hatch from goose barnacles. The only link was that both animals shared similar colouring and are common sights near coasts at different times of the year. Albertus Magnus, one of the few truly rational scholars of this time, dismissed the idea as absurd. However, the concept was widely accepted until the 1700s, and went hand in hand with another ancient belief known as preformationism.

The Great Chain of Being.

A barnacle goose, at one time thought to hatch from goose barnacles.

1856 Alexander Parkes invents the first plastic, parkesine. It is made from the cellulose extracted from wood.
• A survey of the Himalayas shows that Peak XV is the highest in the world, later to be named as Mount Everest.

1857 Elisha Otis invents the elevator, or lift, making it practical to build multi-story buildings in cities where space is at a premium.

1858 The German chemist August Kekulé suggests that carbon atoms could form "a skeleton" around which other atoms bond, and so founds the subject of organic chemistry.
• The first transatlantic telegraph cable is laid.

1859 Charles Darwin publishes *On the Origin of Species* and sets out the theory of evolution by natural selection and immediately causes controversy.

1860 Robert Bunsen and Gustav Kirchhoff use spectroscopy to identify the metal caesium, which they name after the pale blue colour of its flame.
• Stanislao Cannizzaro's address to the Karlsruhe Congress, an international meeting of chemists, shows how atomic weights can be accurately deduced by using Avogrado's law. This is the first step in creating a workable periodic table of elements.

New Materials

Steel, an alloy of iron and carbon, was not invented in 1855. This tough metal had been used for centuries. However, Henry Bessemer's process of producing it made it possible to mass produce high-quality steel reducing the cost considerably. Bessemer mixed precise amounts of iron and carbon – and any other components in a converter. Air was blown through the molten mixture to burn away impurities, and then the liquid steel was simply poured out ready for milling.

The following year saw the arrival of another material. In 1856, parkesine, as it was named by its inventor Alexander Parkes, was the first plastic. It was a mouldable polymer made from cellulose. Highly flammable and brittle, it did not have much use, but the other artificial polymers that would follow it would, along with steel (and aluminium, once made a viable product in 1880s) make the modern world.

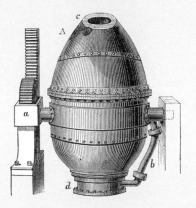
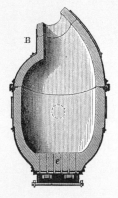

The Bessemer converter.

A, Front view, showing the mouth, *c*; B, Section.

READY TO GROW

Preformationists believed that new bodies grew from eggs – not just the shelled variety but eggs of all shapes and sizes. Their contention was that the egg contained a fully formed, but minuscule, version of the body. The 15th-century Dutch biologist Jan Swammerdam took evidence from the metamorphosis of insects such as butterflies. He insisted that adult butterflies were preformed inside caterpillars, and the caterpillars were preformed in eggs.

In the 1670s Antonie van Leeuwenhoek, the inventor of the microscope and founding figure of microbiology, used his device to observe his own sperm – and that of other animals. On careful inspection, he announced that he could see evidence of a homunculus, or little man. This created an awkward disconnect because until then preformation placed all proto-creatures in the eggs, or ova, of a female, not the sperm of a male. Not every scientist was chauvinistic enough to simply accept this switch, and the school of preformationism rent asunder into the ovists and spermists.

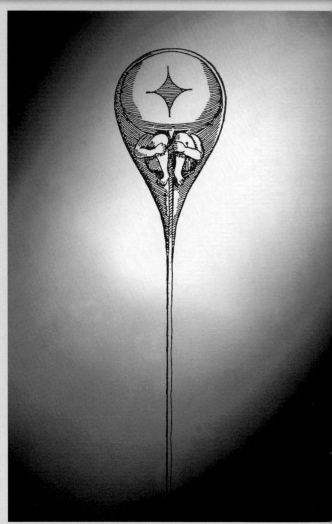

A homunculus image inside a sperm.

Nevertheless, both groups still whole-heartedly opposed the epigenesists. This lesser but just as old set of ideas, once championed by Aristotle, proposed that a new body was created from a set of instructions supplied by every body part. Although it is highly incomplete, this way of thinking is closer to our modern understanding of genetic inheritance, which took shape in the early 20th century. Before that could happen, however, a new evidence-based way of understanding life was needed. It arrived in the mid 1800s to revolutionize biology, finally giving it a scientific foundation.

The revolution resulted in cell theory, which was put forward in 1839 after the work of Theodor Schwann and Matthias Schleiden, two Germans working independently. It had been a long time coming. Biology is an emergent phenomenon, meaning it is an entire level of organization that emerges from the properties of chemistry and physics. It is often said that the more one discovers about physics, and to a lesser degree chemistry, the more one understands about the subject. The same is not true of biology, in that continued observations result in an ever-extending record of unique features, such as body shapes and behaviours. The task of the biologist is to find common themes and overarching phenomena among the hubbub of facts.

Schwann's long years of research showed him that all animals were built from cells or the products of cells, and Schleiden found the same was true of plants. This

Theodor Schwann.

Matthias Schleiden.

was enough to set out the first two tenets of cell theory: Firstly, the cell is the most basic unit of life. Secondly all living things are composed of one or more cells.

Nevertheless, the cells that Schwann, a zoologist, and Schleiden, a botanist, saw were by no means the same kinds of structures. Plant cells were surrounded by a rigid wall made from cellulose. This material is what gives plants their rigidity and structure. Animal cells did not have this wall. In fact, it was much harder to delineate between one animal cell and the next because they also lack uniformity of shape. However, both cell types had a nucleus, a central body that had been discovered in 1831 by Robert Brown, the botanist who is more famous for the Brownian motion, which became a phenomenon studied by physicists.

137

SPLITTING CELLS

As microscopy improved it became apparent that the nucleus was involved in the way cells divided in half. Schleiden described how new plant cells developed through a process whereby a dividing wall formed across the middle, splitting one cell in two. In the 1850s Robert Remak described how animal cells divided in a similar way. (The process of cell division was later named mitosis.) However, Remak's discovery was stolen by Rudolf Virchow who used it to interpolate a third tenet of cell theory in 1855: all cells arise from other pre-existing cells.

This third part of the theory killed off the idea of spontaneous generation for good. (The evidence for that old theory offered by such things as maggots emerging from rotten foods were steadily debunked as the life cycles of organisms were figured out.) Cell theory made it clear that every new body arose from a single cell created by the parents (or parent). However, the question remained regarding how information about growing a body travelled from parents to offspring.

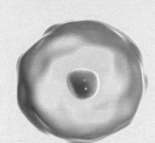

Division of a cell in mitosis.

THE START OF GENETICS

The connection with inheritance was obvious. Traits of all kinds run in families, and farmers had made use of this fact for many centuries to breed better crop plants and livestock. In 1856, an Austrian monk called Gregor Mendel began what would be a research project into the mechanism of inheritance. It would take eight years and involve growing 29,000 pea plants.

By carefully controlling which plants bred with each other, he was able to show how particular features, like flower colour and plant height, were passed on. Mendel found that a plant's inherited traits were due to it receiving a factor for each trait from each parent – resulting in a double set of factors. (Now we know more about the physical mechanism of this process, we have renamed Mendel's factors as genes.)

Mendel encapsulated his discoveries in three laws. The first law was the Law of Segregation, which said that when factors were passed on, the double set was always segregated into two single sets of factors. The second law was the Law of Independent Assortment, which said that the segregation of the first law is random, and each factor is passed on independently of the others. In other words, a gene for flower colour does not habitually travel with a gene for stem length. The final law was the most revealing: the Law of Dominance. A factor (or gene) can exist in several forms. (Today biologists use the term "allele" for the different versions of a gene. For example, the gene for eye colour has the alleles brown, blue and green.) Mendel recognized that the factor for stem length had a tall and a short version

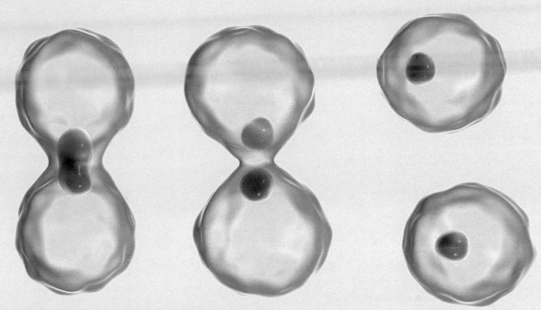

– and flower colour, fruit shape, etc., all had different options. These versions create the great variety seen in the living world which means no two organisms are alike. Mendel's law explains that one version (tall) is dominant over the other (short), which is described as recessive. To grow tall, a plant must inherit just one tall factor. The other one may also be tall or it may be short – the result will be the same. To be short, the plant must inherit two short factors. Therefore dominant traits are more common, but nevertheless recessive traits can (and often do) appear in the offspring of parents with dominant traits.

Gregor Mendel.

139

NATURAL SELECTION

Mendel's work was eventually published in 1865, and promptly ignored for the next 40 years. It was of course already too late to influence the thinking of the greatest biologist of the time. Charles Darwin had published his book *On the Origin of Species* in 1859, after two decades of ruminating over its subject. Darwin's global travels aboard HMS *Beagle* in the 1830s had shown how animals that were obviously related were nevertheless living in very different ways in different habitats. Inspired by the biogeography of Alexander Humbolt, palaeontology of Georges Cuvier, and geological works of Charles Lyell, Darwin began to wonder how life could mould itself to meet the many different challenges of survival.

The result was a process called evolution by natural selection. Frequently cited as one of the most powerful ideas in all of science,

SURVIVAL OF THE FITTEST

The process of natural selection relied on two things. Firstly, no two organisms were identical, with even closely related creatures exhibiting a variety of characteristics. Secondly, life was short and brutal. Death was always an imminent possibility, and only the strongest, best equipped and luckiest individuals survived. This is where natural selection came in. Not all members of a species are up to the particular challenges of their habitat. Only some can find the food, water, and shelter they need while staying away from danger. The ones that are not fit for this task will die, or at least are so outcompeted by their better adapted neighbours that they have fewer opportunities to breed – if any. Nature has selected the fittest members of the population, and deselected the ones unsuited to life in this habitat.

Fit individuals have more offspring, and will pass on their beneficial traits to at least some of them. As this process continues, the proportion of individuals with these beneficial traits increases. Eventually every member of the species has these traits – evolution by natural selection has occurred.

The best illustration of this process comes from the Galapagos finches, 15 species of songbird that live on the isolated Pacific archipelago visited by Darwin in 1835. Each species has adapted to live in a different way, some eating seeds of different sizes while others dig up insects. Upon his return to England, Darwin realized that these 15 species all evolved from one common ancestor that arrived on the islands many generations before. He began to see all life as an immense network, where all organisms, living and extinct, are connected together by a series of common ancestors. The work of piecing together that Tree of Life continues to this day.

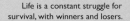
Life is a constant struggle for survival, with winners and losers.

Darwin was able to explain how life is able to adapt to exploit a niche or living space in a particular habitat. The process is very slow, with change taking place over many thousands of generations. However, Lyell's geology showed that Earth was old enough for such a process to occur. It also meant different species occupied the world at different times, with the extinct ones only known through their fossils, as Cuvier explained. And Humboldt's biogeography revealed how the world was filled with particular types of habitat, such as grasslands, deserts and jungle, and in many cases the wildlife in one grassland, let's say, was very different to the residents of another.

One of the most satisfying features of science is when phenomena from disparate areas of research are shown to be linked, with each one revealing something fundamental about the other. This is best seen in the field of electromagnetism, where initially it was found that electricity and magnetism were linked, bringing forth the ability to make motion from a current and a current from motion. Then in the 1860s it was shown that the field of electromagnetism also included the phenomenon of light. This in turn would eventually rock physics to its core by revealing the existence of invisible waves, unlocking the structure of the atom, revealing the diversity of the stars, and most famously of all showing that Newton was wrong about gravity.

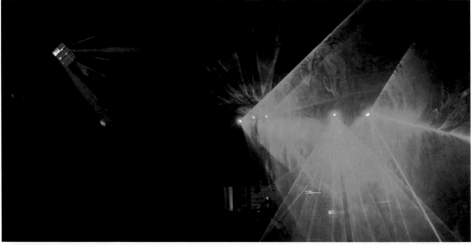

It was in the 1860s that light was included into the field of electromagnetism.

COLOUR CONUNDRUM

The story begins back in Bavaria (now part of Germany) in 1814, when the lens-maker Joseph Fraunhofer developed his precision spectroscope. This was a device that could split white light so exactly, without the blurring and aberrations of earlier lenses, that the observers could look at the coloured spectrum of light in great detail through a magnifying eyepiece. They could clearly see razor-thin gaps, dark lines among the colour. These so-called Fraunhofer lines were visible in the light from the sun, but no one was really clear about what they were. By contrast, the spectroscope showed that light from a flame comprised a handful of very distinct

Bunsen burner.

Flame test for potassium

Flame test for sodium

1861 William Crookes identifies thallium using a spectroscope. He names the heavy metal after the leaf green colour of its emission spectrum. Gustav Kirchhoff and Robert Bunsen use the same technique to discover rubidium, which they name after the deep red from its ▓▓▓▓▓▓▓

1862 Battleships protected with thick iron armour engage in the first ironclad battle at Hampton Roads during the American Civil War. The engagement is a draw.

1863 London's Metropolitan Railway becomes the world's first underground mass transit system, or subway. The steam-powered trains run in tunnels cut below roadways.

1864 James Clerk Maxwell publishes his unifying theory of electromagnetic fields, which shows that light is an electromagnetic wave.

colours, not a rainbow-like spectrum. Why the difference?

The flame test was an analytical technique used by chemists that had been inherited from the days of alchemy. When a substance is burned, it produces a flame with a distinctive colour. This had been well illustrated by Humphry Davy upon his discovering of potassium in 1807. This soft, silver metal burned with a unique lilac flame. In the early 1850s, Robert Bunsen, a chemist working in Heidelberg, Germany, developed a means to measure the colour of flames. For that he needed a source of heat that would not interfere with the colours of burning material. The result was the Bunsen burner, a now ubiquitous laboratory tool. The burner is fuelled by gas, and a collar at its base allows the user to control how much air mixes with the fuel. When air is restricted, the burner produces a bright yellow flame, and by adding air, the flame fades into a hot blue cone. In this set-up, the gas fuel is burning very fast and hot, and so not interfering with the light coming from other flames.

LAWS OF SPECTROSCOPY

Towards the 1850s, Bunsen began to collaborate with fellow German Gustav Kirchhoff. They used a spectroscope to record the light in the flames of a wide range of materials. They found that the colours present were unique to each material, and so the set of colours, termed as the emission spectrum, could be used as a very accurate way of determining the presence of elements in a compound. Indeed, the pair managed to discover elements that were hitherto unknown. They named both metals after the most distinctive colours in their emission spectrum: rubidium for red and caesium for sky blue.

However, Kirchhoff was able to take this phenomenon a step further and in doing so explained the dark Fraunhofer lines. The result was a set of three laws of spectroscopy.

1   A hot, incandescent material gives out a continuous emission spectrum. This is what the Sun and a star are doing.
2   A hot gas will produce an emission spectrum of distinct coloured lines, not a full rainbow.
3   When a full spectrum shines through a cold gas, some of the colours are absorbed. Like an emission spectrum, this absorption spectrum is unique to each element.

This final law explained the mysterious Fraunhofer lines. The gases that surround the Sun are absorbing some of the light from the hot surface, creating dark empty lines in the continuous spectrum. Therefore, it was possible to identify the elements in a star's atmosphere just by looking at the absorption lines in its light. This fact soon showed that the elements found down here on Earth were also present in all parts of the Universe – and there were even some unknown ones too (see box *opposite*).

A spectroscope splits the coloured light from a flame.

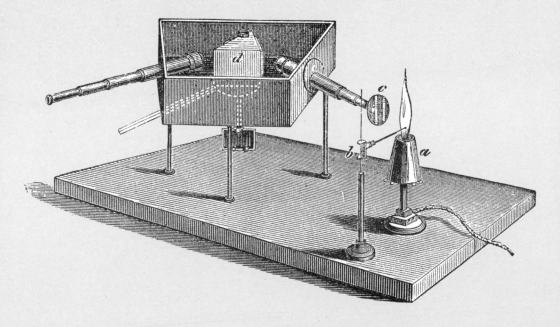

1865 The concept of entropy is proposed by Rudolf Clausius to complete the second law of thermodynamics, which states that systems move from low entropy to high entropy – i.e. hot things always transfer energy to cold things. The direction of energy is driven by random chance.

• Gregor Mendel sets out his three laws of inheritance. They are largely ignored but become important in the early 20th century as the field of genetics is developed.

1866 Alfred Nobel invents dynamite, a form of nitroglycerine explosive that is stabilized and made safe to handle by mixing it with a very fine earth.

1867 The Covington–Cincinnati Suspension Bridge is built across the Ohio River using a design that is enlarged and improved by its architect John A. Roebling to build the Brooklyn Bridge. The Ohio bridge was renamed for Roebling in 1983.

1868 Tell-tale wavelengths in light coming from the Sun reveal an unknown element. The new substance is named helium.

SUN GAS

On 18 August 1868, astronomers gathered in southern Asia to train their spectroscopes at the Sun. That day marked a total solar eclipse, and as the Sun's bright disk was blocked out for a few minutes, the astronomers would get a very clear view of the corona, the fainter sphere of gas that surrounds the star. A French astronomer, Pierre Jansen, saw some unknown emission lines in the coronal light. A few months later, the English astronomer Norman Lockyer saw the same thing in the solar spectrum, and concluded that he was seeing evidence of an element that was unknown on Earth. Assuming it was metal, Lockyer named it helium, meaning "sun metal." In 1895, the Scottish chemist William Ramsay collected pure helium from a mineral. It turned out to be a gas, and a very strange inert material at that. Today, helium belongs to a group of elements called the noble gases, which are so-called because they are so chemically inactive that they never interact with the "common" elements.

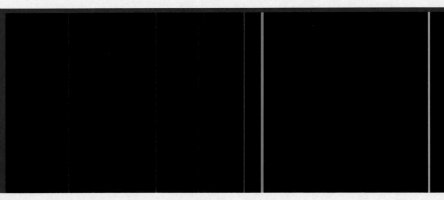

Helium emission spectrum.

GLOWING TUBES

Around the same time as Bunsen and Kirchhoff were systematically recording the emission spectra of flames, a pair of researchers in Bonn had found another way of seeing the same kinds of colours. Heinrich Geissler was asked by Julius Plücker to make a glass tube with as complete a vacuum as he could manage. Plücker also asked him to add electrodes so he could pass electric current through the nothingness. Like Hauskbee and his glass sphere before him, Plücker found that the electrified tube glowed in the dark. More importantly he could make it glow with a variety of colours by adding a faint trace of different elements, and these colours matched the emission spectra recorded by Bunsen. (The Geissler tube was the forebear of the neon light, so named because neon gas gives out a dark red. Today these kinds of light sources are better known as gas-discharge lamps. The most common form is the low-energy light bulb, where the light is produced using minute amounts of mercury vapour.)

So it appeared that there was some kind of unified property of each element that produced a certain kind of light when it was given energy, either by being burned or electrified. Plücker also found that the light in his tubes could be attracted and repelled by magnets.

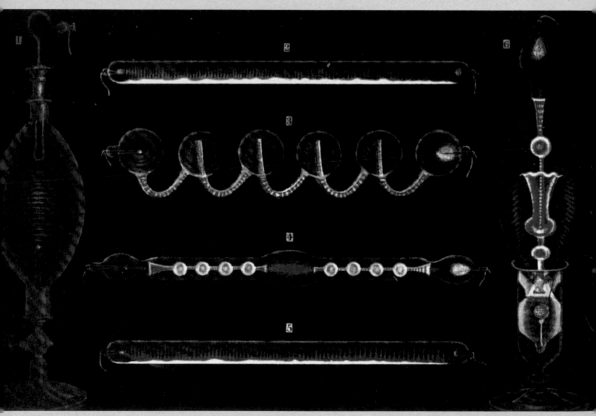

Geissler tube.

FIELD EQUATIONS

The man who unified all these observations was James Clerk Maxwell, often called the Scottish Einstein. (Indeed, it was Maxwell's work on explaining electromagnetism that got Einstein on the path to developing his theory of relativity.) In the 1850s Maxwell took over from Michael Faraday as the leading figure in electromagnetism (Faraday had retired by this point). Maxwell spent most of the 1850s studying what Faraday called the "lines of force" that acted around electricity and magnets. Today we would know them better as a force field. Maxwell was able to show that the force fields changed their strength at the speed of light, which suggested that all three — electricity, magnetism and light — were part of the same phenomenon. By 1864 Maxwell had unified all three into electromagnetic fields by producing a series of equations that explained how their characteristics were related.

Maxwell's work showed that light was a form of electromagnetic radiation that carried energy as oscillations in an

James Clerk Maxwell.

electromagnetic field that permeates the Universe. This is why light can travel through a vacuum and outer space, while other waves, such as sound, only work in a medium such as air or water.

The colour of light is a feature created by our sense of vision in response to the eye detecting light with different wavelengths. Red light has a longer wavelength (and lower energy) than yellow, then green and finally blue and violet light, which have the shortest wavelengths. William Herschell had already described infrared, or invisible heat radiation. This has a longer wavelength than red light. Ultraviolet had also been found at the other end of the spectrum by Johann Ritter, again an invisible ray but this time with more energy than visible light. Maxwell predicted that the electromagnetic spectrum would contain more invisible forms. The search for them began.

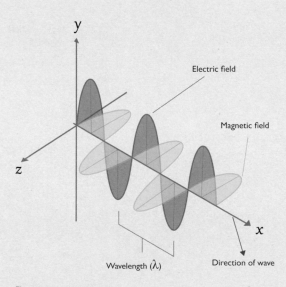

Electromagnetic wave.

The periodic table is one of the most powerful tools in science. In its current form it contains 118 elements, the full set of building blocks that make up the Universe — although the last 25 or so are entirely artificial. At a glance, the trained eye can glean some insight into the properties of any element from its position in the table relative to its neighbours. The trends of shared and contrasting characteristics that the periodic table presents are due to the subatomic structure of the elements. It is therefore all the more remarkable that the table was put together 30 years before the first subatomic particles were even identified.

Group→	1	2	3	4	5	6	7	8	9	10	11	12	13	14	15	16	17	18
Period 1	1 H																	2 He
2	3 Li	4 Be											5 B	6 C	7 N	8 O	9 F	10 Ne
3	11 Na	12 Mg											13 Al	14 Si	15 P	16 S	17 Cl	18 Ar
4	19 K	20 Ca	21 Sc	22 Ti	23 V	24 Cr	25 Mn	26 Fe	27 Co	28 Ni	29 Cu	30 Zn	31 Ga	32 Ge	33 As	34 Se	35 Br	36 Kr
5	37 Rb	38 Sr	39 Y	40 Zr	41 Nb	42 Mo	43 Tc	44 Ru	45 Rh	46 Pd	47 Ag	48 Cd	49 In	50 Sn	51 Sb	52 Te	53 I	54 Xe
6	55 Cs	56 Ba	57 La	* 72 Hf	73 Ta	74 W	75 Re	76 Os	77 Ir	78 Pt	79 Au	80 Hg	81 Tl	82 Pb	83 Bi	84 Po	85 At	86 Rn
7	55 Fr	88 Ra	89 Ac	* 104 Rf	105 Db	106 Sg	107 Bh	108 Hs	109 Mt	110 Df	111 Rg	112 Cn	113 Nh	114 Fl	115 Mc	116 Lv	117 Ts	118 Og

	*	58 Ce	59 Pr	60 Nd	61 Pm	62 Sm	63 Eu	64 Gd	65 Tb	66 Dy	67 Ho	68 Er	69 Tm	70 Yb	71 Lu
	*	90 Th	91 Pa	92 U	93 Np	94 Pu	95 Am	96 Cm	97 Bk	98 Cf	99 Es	100 Fm	101 Md	102 No	103 Lr

The periodic table of elements.

LOOKING FOR ORDER

Scientists have always sought ways to classify the elements. For example, the four classical elements were seen as a continuum of heat and moisture, with fire being hot and dry, air being hot and wet, water being cold and wet and earth being cold and dry. After many alchemical blind alleys, the journey to the periodic table began with John Dalton's atomic theory. Dalton began to unpick fundamental characteristics of elements, which presumably were due to the differences in their atoms. His law of definite proportions told him that certain atoms could bond to differing numbers of atoms. In turn that gave Dalton the insight that it could be possible to ascribe a weight to the atoms of each element. However, he lacked the information needed to figure out the exact proportions in molecules or to show that atomic weights are a unique feature of an element rather than simply a variable.

In 1811, the Italian scientist Amedeo Avogadro proposed a way of breaking this deadlock. He said that any gas at a constant pressure, a constant temperature and in a fixed volume would always have the same number of particles — be they atoms or more likely molecules — as any other gas in the same state. This would open the door to measuring the relative weights of elements.

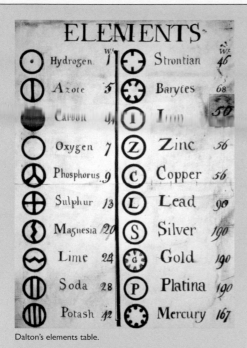

Dalton's elements table.

Amedeo Avogadro.

1869 Dmitri Mendeleev proposes the periodic table. This lists elements in order of atomic weight and organizes sets of elements in rows according to a repeating, or periodic, pattern seen in their valence, or combining power.
• The Transcontinental Railroad is completed across the United States with competing tracks meeting in Utah.
• A chemical identified as deoxyribonucleic acid, or DNA, is identified in the pus collected from wounds by Friedrich Miescher.

1870 Construction of the Brooklyn Bridge begins. (It is not completed until 1883.)
• The Weather Bureau, the United States' national meteorological service, is founded and issues its first weather forecast.

1871 A telegraph cable running all the way from London to Darwin, Australia is connected.

1872 Yellowstone National Park becomes the first national reserve set up to protect wildlife and natural landscapes.

1873 Willoughby Smith discovers that selenium produces electric currents when illuminated by light. This property forms the basis for early television technology.
• Carl von Linde begins building industrial-scale refrigeration.

1874 Ferdinand Braun discovers the phenomenon of semiconduction, where a material can be made to switch between an electrical conductor and insulator. This property will form the basis of electronic technology.

1875 Gallium metal is observed by spectroscopy. When it is purified it matches the properties of eka-aluminium as predicted by Dmitri Mendeleev in 1871.

1876 Alexander Graham Bell and Elisha Gray both patent their independently developed telephone systems.

For example, a jar of hydrogen gas weighed 16 times less than a jar of oxygen. Hydrogen is the lightest element and so it was given an atomic weight of 1, and in turn oxygen had an atomic weight of 16. However, Avogadro's suggestion was mostly ignored by the scientific community, and dismissed by those who had read it.

THE PERIODIC TABLE

GETTING ORGANIZED

In the 1820s, Jöns Jacob Berzelius, a Swedish chemist with a zeal for organization, began the practice of giving elements symbols composed of letters. H was for hydrogen, O for oxygen, and so on. To break down language barriers, some symbols were derived from the Latin names for elements. Hence iron has the symbol Fe for *ferrum*, while lead is Pb for *plumbum*. The convention started by Berzelius was soon well established. New elements were given a symbol with one or two letters only, with the first always in upper case, the second always in lower case. Today new elements are assigned temporary three-letter symbols until an international agreement on a name can be reached.

Jöns Jacob Berzelius.

Berzelius also introduced the idea of chemical formulae. Water had been shown to be two parts hydrogen for every one part water, and so Berzelius represented this as H_2O. Eagle-eyed readers will see that this initial notation has since been changed to H_2O.

VALENCE

A clearer understanding of the way elements combined in compounds arrived in the 1850s. After studying under Robert Bunsen (who had yet to become the famous figure he is now) the English chemist Edward Frankland began research into the reactivity of metals. He found that the metals always combined with other reactants in the same proportions, no matter what they were reacting with. Just as importantly, the proportions were not the same for every metal. He described this property as "combining power", but it is better known today as "valency", derived from the Latin word for strength.

Frankland's work created a firm foundation for a theory of valency, where once it was only vague ideas. Soon every atom was being ascribed with a value for it. Hydrogen has a valence of one, while oxygen's was two. In other words, hydrogen atoms can only bond with one other atom at a time. Oxygen can link

Edward Frankland.

up with two, as it does in water (H_2O). Carbon has a valence of four.

In 1858, Friedrich August Kekulé showed that carbon's ability to bond with up to four other atoms, including other carbons, made it possible for this element to form molecules that were immensely long chains, branching networks and even

Friedrich August Kekulé von Stradonitz, German organic chemist.

1877 Giovanni Schiaparelli draws a map of Mars and some copies translate his depiction of channels as "canals". This appears to confirm the belief of many that Mars is home to aliens.

1878 Eadweard Muybridge produces the first moving image, or movie, using a series of still images of a horse galloping.

1879 After 20 years of research, Louis Pasteur completes his germ theory of disease which proposes that disease is caused by microorganisms.

1880 Joseph Swan in England and Thomas Edison in the United States independently invent the incandescent light bulb.

rings. Chemicals made by life – organic entities – are always based on carbon, and the term organic chemistry has now been extended to mean just about any chemical with carbon in it. Of the 10 million molecules described by chemists, more than 90 per cent are organic in nature.

All thanks to Kekulé's vision of carbon's combining power – which came to him in a dream while asleep on a bus in south London!

MARTIAN CANALS

In 1877, the Italian astronomer Giovanni Schiaparelli drew up a map of the surface of Mars, taking full advantage of the once-in-a-generation proximity between Earth and the red planet. His map was filled with what he saw as geological structures like plains and plateaux with "canali" running across the surface. Although that word means "channel" in Italian, it was translated as "canal" on English versions of the map. At once the Anglo-Saxon world leapt at the idea that Mars was home to some very industrious aliens. The race began to spot these neighbours, who few thought to be friendly. Leading the pack was the industrialist Percival Lowell, who built an entire observatory on Mars Hill above Flagstaff, Arizona. His staff found no aliens, but they did find the first crucial clues of the expanding Universe and spotted Pluto (almost called Percival) which was declared to be the ninth planet (for a while at least). No one can see Schiaparelli's channels on Mars today. It appears that what he saw were the blood vessels of his own retina reflected in the eyepiece!

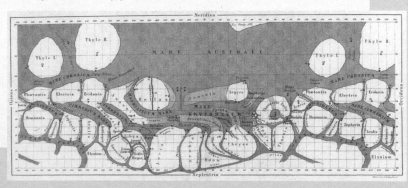

Giovanni Schiaparelli's map of Mars.

KARLSRUHE CONGRESS

In 1860, Kekulé was one of the conveners of an international gathering of chemists at Karlsruhe, Germany. The Karlsruhe Congress was an opportunity for experts to exchange ideas and find new areas for research. One of the speakers was the Italian Stanislaus Cannizzaro, who gave a talk on the work of Amedeo Avogadro, which was met with much greater interest due to the 50 years of discovery that had occurred in the interim.

Avogadro's method, eventually recast as Avogadro's law, allowed chemists to calculate the relative atomic masses of the elements – of which there were 59 at the time. The next logical step was to order elements according to their weights, and soon people began to see patterns. For example, in 1865 the English chemist John Newlands observed that the valence and atomic weights formed natural groups of seven. Newlands called this fact the "Law of Octaves" and even organized the elements like notes on a stave to illustrate the way the properties changed steadily from the first to the seventh member, and then reverted back to the start with the eighth in line forming the first member of the next group. He ran the system all the way to thallium, a heavy metal, but could not keep the rhythm of properties going.

PERIODIC PROPERTIES

Instead of a musical score, a Russian chemist tried to make sense of the elements – by now fast approaching 70 in number – while playing a game of solitaire, the solo card game. He was trying to figure out how to stack up the

Early Mendeleev periodic table.

elements according to their characteristics like Newlands. He characterized the same sets of elements as not octaves but periods; it was called that because their properties were repeated, or were periodic.

This chemist was Dmitri Mendeleev, perhaps the most famous chemist in history (there is not much competition). He too hit problems, because he was working with an incomplete set of elements and was uncertain whether to fill gaps with guesswork or leave them empty. In the end he did a bit of both, creating a side group with metals like gold, copper and nickel that did not seem to fit anywhere. When assembled, the periods became rows, like the cards in solitaire, and they formed up into columns, which Mendeleev called groups. The elements in a period increased in atomic weight, and

also rose in valence from one to four and then back down to one again. Then a new period started and repeated the pattern. Members of a group had the same valence — one for Group 1, two for Group 2 etc. — but they were widely separated by weight.

Mendeleev presented the first periodic table in 1869, but spent decades in refining it with better information. From the start, however, he was able to predict the atomic weights and valence of the as-yet undiscovered elements that would fill the gaps. On top of that, he estimated their density, boiling and melting points, and even colours. Time and again these predictions were proven true as the periodic table of elements was steadily filled. It would take the birth of quantum physics to explain why the table worked so well, but work it did.

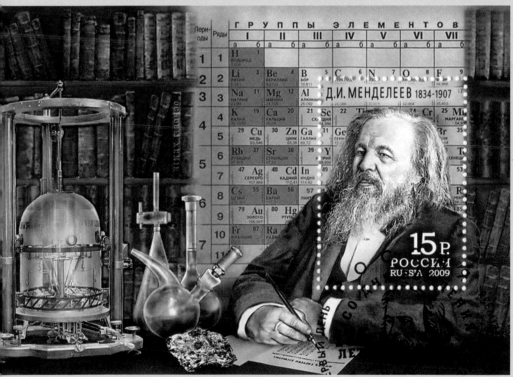

Dmitri Mendeleev.

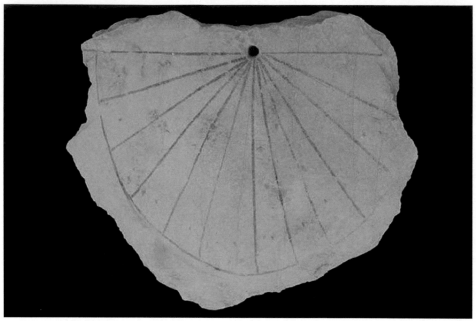

We have always needed to measure time. This Ancient Egyptian sundial from around 1500 BCE divides the day into 12.

Time and, more importantly, the measurement of time, is central to many parts of everyday life. We rely on timepieces to count out a standard rhythm to orchestrate our daily schedules – and take it for granted that our timings are consistent with everyone else's. This is all thanks to a global standard of time which has been in place since 1884, and which has been built out of discoveries about the Solar System, the Earth, motion, and energy.

WHAT IS TIME?

Time is one of the thorniest subjects of science. We all know what it is, but no one can really explain it. Time is often said to be the fourth dimension along with the three dimensions of space: length, width and depth. Those three dimensions describe the size and shape of an object and its location in space. Time documents how the object changes its shape and location. As well as having a place in space, all things have a location in time.

In the 1850s, Rudolf Clausius introduced the idea of entropy to describe the behaviour of energy. His whole concept became known as the second law of thermodynamics, which states that the entropy of a system tends to increase. By "system" we could mean anything from a generic

Rudolf Clausius.

1882 The Edison Company sets up the world's first power stations to provide electricity to paying customers in New York City and London.

1883 Construction of the Brooklyn Bridge across the East River in New York is completed.

1884 Sandford Fleming calls the International Meridian Conference in Washington, DC, to standardize the world's time zones under the Coordinated Universal Time (UTC) system.

1884 The 10-storey Home Insurance Building is built in Chicago. It has an internal steel framework which makes it the world's first skyscraper.

1885 Karl Benz invents the Benz Patent-Motorwagen, a three-wheeled vehicle powered by an internal combustion engine, which becomes the first practical car.

1886 Pure fluorine, a gaseous element so reactive it reacts with almost everything, is finally isolated by Henri Moissan 90 years after it is predicted to exist by Antoine Lavoisier.
• Clemens Winkler discovers germanium, predicted to exist by Mendeleev in 1871.

1886 Nikola Tesla, a Serbian engineer and inventor working in the United States, develops the transformation system that controls the voltage of alternating current used today in modern electricity networks.
• Martinus Beijerinck discovers that some bacteria can take nitrogen from the air and add it to soil. This is the first step in understanding the nitrogen cycle.

1887 Ernst Mach reveals that objects, such as bullets, can break the sound barrier.
• The Michelson–Morley experiment attempts to show that empty space is filled with a universal substance called ether. The theory says that speed of light waves varies by minute amounts due to the motion of Earth through ether, but the results prove that this is not the case. Ether does not exists.

1887 Radio waves are discovered by Heinrich Hertz using a simple transmitter that creates a signal using an electrical spark.

1888 Theodor Boveri shows that chromosomes are involved in genetic inheritance.

1889 Activation energy, the boost needed for chemical reactions to proceed, is described by Svante Arrhenius.

object, such as a teapot, to a star or entire galaxy. Increasing in entropy means its disorder increases, which is another way of saying the energy in the system becomes less concentrated.

So, a just-filled teapot starts out as being hot. It has low entropy because there is a lot of heat concentrated inside. Gradually but inexorably that heat will spread, thus it makes the surroundings warmer and the teapot colder. Eventually the tea inside is at the same temperature as the surroundings. This is not because heat (or energy) has stopped moving, it is just that heat moving out of the teapot is in equilibrium with the heat moving into it, so overall there is no change in temperature, no change in entropy. This goes to the heart of entropy as being the result of a random, unpredictable process. Energy has no direction, it is just that statistically it is more likely to move from areas of high energy to areas of low energy. It does not break any of the laws of thermodynamics for heat to move from the surroundings in to the teapot making it even hotter, but it is so unlikely that it never happens.

Hot tea always cools down, never warms up.

THE ARROW OF TIME

Just by chance alone, energy spreads out. Eventually the teapot will crumble to dust, a star will fade away, and a galaxy will dissipate as a cloud of dust and gas. As far as anyone can tell, this increase in entropy is what gives time its direction. Time always goes forward because it results from the changes in systems (the teapot or galaxy) caused by entropy, and entropy never stops increasing. That's why we cannot make time go the other way so we all get younger.

In the 1900s, Einstein wrapped time and space together into a new model of the Universe, and today's scientists are still pondering what that tells us about time. However, in ancient days our interests in it were focused more on matters of life, death and the afterlife. Clocks and calendars were not invented by scientists but by farmers and priests. Farmers were interested in the best times of the year to reap and sow. They counted days and watched the stars as they made a full circuit of the heavens every year. Often crucial days in the agricultural

The astrolabe could be used to tell the time from the position of the stars.

Clockwork.

calendar held religious importance too, but priests also wanted to know what time of day it was so they could perform rituals and pray at the right times. The day was divided into handy units called hours – 10 for the hours of daylight and one either end for dawn and dusk, making 12 hours in a day (and another 12 at night.) For added accuracy, the hour was divided into a much smaller, or minute, subunit – the minute. In turn, the minute was divided a second time into smaller units still – the second minute, or just second. (These divisions are in 60s because that is how the ancient Babylonians counted – and it works so well we have not changed it.)

DAY TO DAY

Days had a clear start and end from sunrise to sunset but the length of time between them varied with the seasons. The only fixed point was noon, where the Sun was at its highest point in the sky in the exact middle of the day. Horologists, or time experts, were busy each day observing the position of the Sun, and using its highest point to recalibrate their clocks.

Copernicus proved that the Sun's progress through the sky is the result of the rotation of the Earth from west to east, and so the time of noon varies according to where the observer is. People in the east see noon earlier than those to the west. For example, noon in the western English city of Bristol is about 10 minutes after noon in London. That posed no practical problems until railways were built between the two port cities in the 1830s. The noon train from London travelled for two hours to reach Bristol, and arrived at ten minutes to two in the afternoon by local time. To stop all the confusion, the railways adopted a standard time, set by the observatory at Greenwich. Noon there was noon at every station in the land.

A falling red ball was once used to mark noon at Greenwich Observatory in London.

TIME ZONES

It is not practical to impose a standard time over a large area. Every 15 degrees of Earth's circumference results in a one-hour difference in the local noon. In 1883, the United States was divided into four time zones, roughly following the 15-degree rule. The following year at a conference in Washington, DC, the world powers (mostly) agreed to base the world's time on Greenwich Mean Time – "mean" roughly means "average" – with 24 time zones taking their cue from London. Today we still organize time zones on this system, although the official time now comes out of Denver, Colorado, not east London.

As the last decade of the 19th century approached, physics was increasingly becoming the study of a selection of invisible phenomena that seemed to permeate through space in a mysterious way. The anchor point for this research was James Clerk Maxwell's theories on electromagnetic radiation, and in general the invisible mysteries were all treated as some kind of ray, wave, or radiation. Physicists would eventually find that they were investigating a range

THE CATHODE RAY

The first invisible phenomenon in the story could actually be seen once you turned the lights out. In the early 1870s, English physicist William Crookes redesigned the vacuum tube so it had 10,000 times less gas than similar devices built by Geissler 20 years before. It was also able to apply much greater voltages between two electrodes — the negative cathode and positive anode — which stood within the sausage-shaped glassware.

In a darkened room, the "Crookes tube" did not glow in the same way as Geissler's. Instead area close to the cathode

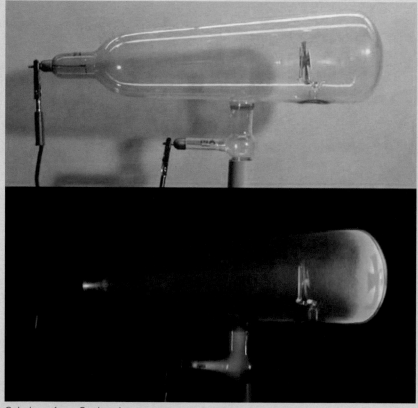

Cathode rays from a Crookes tube.

of phenomena. All emanated from the properties of the atom but had quite distinct characteristics. Nevertheless, the terminology of radiation still permeates them all, leading to frequent confusion and unwarranted concern.

1890 The mitochondrion is described by Richard Altmann. Later this is found to be the organelle that provides a cell with its energy.
• Francis Galton proves that human fingerprints are unique to individuals.
1891 The first hip replacement surgery is attempted with ivory implants used to replace damaged bone.

William Crookes.

remained dark and an eerie green glow appeared to strengthen in intensity as it neared the anode. The light moved past the anode, shining out of the tube. The light had a definite direction, always running from the cathode to the anode and not radiating out in all directions. As a result, the phenomenon was called a cathode ray – and Crookes's device became better known as a cathode ray tube. Fifty years later, cathode ray tubes formed the basis of electric televisions, but for now they were the focus of intense research. Eventually they would shine the first light on a new science: subatomic physics, but before that there were more invisible rays to find.

TWO TYPES OF CELL

Microbiology has always been restricted by the power of microscopes, and it was not until the 1880s that these devices were able to resolve structures within cells well enough to differentiate them with confidence into a set of organelles with specific anatomies and functions. In 1890, the German pathologist Richard Altmann reported seeing "bioblasts" which we now know as mitochondria. We know now that these capsule-like structures are the locations where respiration takes place – where the cell oxidizes glucose into carbon dioxide and water and releases energy. The discovery of this crucial organelle, without which none of our cells could exist, was the clearest indication yet of a stark delineation between two types of life form. The eukaryotes have large, complex cells with a nucleus, mitochondria and other organelles. Humans are eukaryotic along with every other animal, plant, fungus, and a suite of unicellular organisms like amoebas and protozoa. Bacteria and their like are prokaryotes with cells hundreds of times smaller than ours, which have no distinct organelles.

SPARK OF DISCOVERY

The source of a cathode rays light did not match with regular thinking about electromagnetic waves. James Clerk

Guglielmo Marconi with radio equipment.

Maxwell's equations described light waves as simultaneous oscillations in the electric and magnetic fields. The speed of all light waves was the same, and so as the frequency increased, the wavelength decreased. Higher frequency waves carried more energy. Maxwell was unsure of an upper limit to the energy in radiation, but felt sure that lower-energy waves existed. He predicted that light could be made using a current of electricity.

In 1887, ten years after Maxwell's death, Heinrich Hertz built a device to detect these invisible waves. The apparatus sent a spark of electricity across a gap between two brass balls. As well as producing a flash of visible light, the device would give out invisible electromagnetic radiation. To detect it, Hertz had a simple receiver, which was a loop of wire with a similar "spark gap" in it, placed across the room. Working mostly in complete darkness, Hertz finally saw a tiny spark picked up by his receiver. Invisible waves were radiating out from transmitter circuit and inducing a very small current in the receiving ring.

For a short while this phenomenon was named "Hertzian waves", but was soon better known as radio – taken from radiation. By

1892 James Dewar invents the vacuum flask which isolates substances inside a vacuum jacket. Heat can only move through this jacket very slowly via radiation and so material in the flask stays cold or hot for long periods.

1893 Rudolph Diesel patents the internal combustion engine design that bears his name.

1894 Lord Rayleigh and William Ramsay isolate pure argon by cooling the air so its constituents liquefy. This is the first sample of an inert noble gas.

1894 Charles Parsons invents the steam turbine engine and showcases its power in *Turbinia*, a high-speed launch. Steam turbines are used in battleships for the next 50 years.

1895 X-rays are discovered by Wilhelm Roentgen.

1896 The Russian maths teacher Konstantin Tsiolkovsky proposes a practical space launch vehicle based on a multistage rocket.

• Henri Becquerel discovers radioactivity from the emissions produced by uranium minerals.

1897 English physicist J. J. Thomson discovers that cathode rays are actually streams of tiny negatively-charged particles which he names electrons. Thomson finds that electrons are hundreds of times smaller than hydrogen atoms, the smallest form of atom.

1901 Italian Guglielmo Marconi had improved radio technology enough to transmit a coded message from England to a receiver in Canada. Hertz was not forgotten: the unit of frequency, or wavelengths per second, which is a crucial factor in radio technology (as well as fundamental to all waves) was made the hertz (Hz). Radio waves have a frequency measured cenitimetres, metres and even kilometres. The next discovery in invisible rays was to be at the other end of the spectrum – literally.

Heinrich Hertz.

Heinrich Hertz's oscillator.

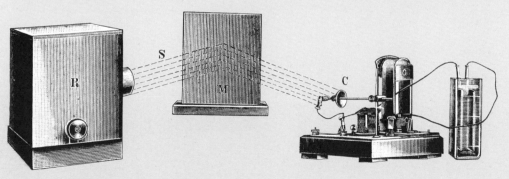

X-RAY VISION

The next radiation discovery was a complete accident. In the 1895, the German physicist Wilhelm Roentgen was prepared to use his cathode-ray tube, by now even more advanced and powerful than Crookes's original. The delicate tube still had its cover on but Roentgen had switched it on and was preparing a target of light-sensitive paper. The paper became fogged by an invisible beam, which Roentgen recorded with a quizzical "X". In due course, Roentgen named his discovery X-rays, and his investigations revealed they could pass through some solid objects. He used the rays to photograph the bones in his wife's hand, who is reported to have

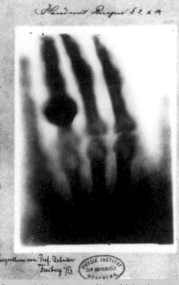

Roentgen's X-ray picture of his wife's hand.

A New Pathogen

In 1898, the Dutch microbiologist Martinus Beijerinck's study of the tobacco mosaic disease, a serious threat to commercial tobacco production, showed that it was caused by an infectious agent smaller than a bacterium. He named it a virus. Beijerinck could not grow viruses in the lab so he concluded that it could only multiply in the living plant. It took until 1941 for X-ray crystallography to reveal the true nature of a virus: a strand of DNA coated in a protective protein. The agent replicates by hijacking the machinery of a cell, creating so many copies of itself that the cell ruptures – which is the root of viral disease.

Martinus Beijerinck.

been rather disturbed by the now-classic negative image of ghostly white bones shadowed on a black background. During the following years, doctors in Glasgow, Scotland, were using X-ray machines to look for fractures, and in 1898 the British Army took mobile X-ray units for use on the battlefield of Sudan.

No one quite knew it at the time, but the penetrating power of X-rays was due to the immense energy they carried, greater than UV, and this energy posed a danger to life and limb. Very soon an entirely different kind of radiation was discovered which was even more deadly.

1898 Neon, krypton, xenon, and radon gas, the remaining noble gases, are isolated by William Ramsay and Morris Travers.
• Marie and Pierre Curie isolate two new metals among the uranium in radioactive minerals. They name them polonium and radium.
• Martinus Beijerinck discovers the first

1899 Actinium, another radioactive heavy metal, is found in the uranium mineral pitchblende.

1900 Ernest Rutherford and Paul Villard describe radioactive emissions as three types: alpha, beta, and gamma radiation.
• Max Planck finds that objects release radiation in packets called quanta.

PLANCK'S CONSTANT

By the 1890s, it was clear that electromagnetism as described by James Clerk Maxwell was not quite right. The most intense radiation emitted by objects increased in frequency as the temperature increased. The human body, kept at the relatively cool 37°C, emits only infrared, and a red-hot poker or blindingly bright star emit higher frequency visible light because they are much hotter. The maths to describe this suggested that the hot bodies should be blasting out much more energy in the form of UV and X-rays. The German physicist Max Planck was able to match up the maths with the observed radiation by dividing energy into quanta, or packets of specific sizes. Energy did not flow out of objects like water from a jug. Instead it left as a succession of quanta, which had a particular and unchanging magnitude. Quantum physics had arrived.

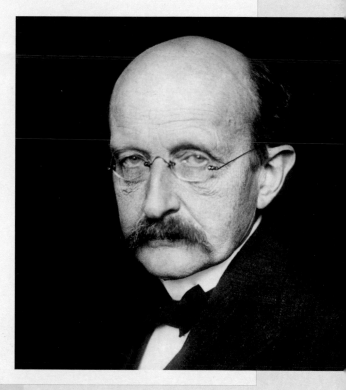

Max Planck.

RADIOACTIVITY

Henri Becquerel.

Some minerals glow in the dark after exposure to bright light. The reasons for this fluorescence and phosphorescence are now understood to be varied and complex, but in 1896 Henri Becquerel, a French physics professor, wondered if their glow was linked to invisible rays like the glow of a cathode-ray tube. He wrapped a wide range of minerals in a paper and placed them in a photographic plate and waited to see whether any invisible emissions would fog the plate. Nothing happened for a long time. But then he tried a sample of uranyl sulfate, a uranium mineral better known today as yellowcake, a precursor material to nuclear fuel. At the time uranium was seen as a harmless heavy metal that gave pottery and glassware a pleasing yellow–green pigment. All that was about to change.

The uranium salt fogged the photographic plate. The Frenchman tried other uranium compounds and found they all gave out "Becquerel rays". A few years later Becquerel's Polish research student, Marie Curie, renamed the phenomenon "radioactivity". She showed that radioactivity arose from certain elements, the most common being

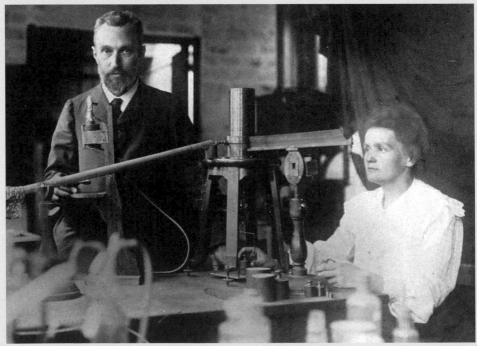

Marie and Pierre Curie.

uranium and thorium. Curie worked tirelessly to extract tiny amounts of other radioactive elements, such as polonium and radium, from uranium minerals.

In 1898, a young New Zealander called Ernest Rutherford, working at Cambridge University, called into question whether Becquerel rays were radiating in the same sense as radio waves, light, and X- rays. Rutherford found that radioactive metal gave out two types of radiation, which he called termed alpha and beta. In 1900, Frenchman Paul Villard found a third type of radiation, which he added as gamma radiation. Only this third type belongs to the electromagnetic spectrum, as the most energetic member. The other two types of radioactivity were soon to be revealed as not waves at all but streams of particles, particles that were even smaller than an atom.

THE ELECTRON

The circle of discovery now returned to the cathode ray, still enigmatic after the revelations of the 1890s. Physicists knew they existed in a vacuum like electromagnetic waves, but unlike them the rays flowed like a gas and interacted with magnetic fields like a metal. The best description offered was "radiant matter".

Heinrich Hertz had found that cathode rays were unaffected by electric fields. In 1897, the English scientist John Joseph "J. J." Thomson reran the experiment, just to check. His apparatus was of a higher specification than Hertz's and he found that the rays interacted with electricity. (Earlier cathode-ray tubes had too much gas left in them for this effect to be seen.)

Thomson showed that the rays swung towards the positive charge, and thus cathode rays were negative. Thomson concluded that cathode rays were not rays at all but a stream of particles. To calculate their size, he compared how the extent to which magnetism deflected the beam and the results were astounding. The charged units in a cathode rays were 1,800 times lighter than hydrogen atoms, the lightest of all atoms. Thomson gave them the name electron, the first of the

J. J. Thomson.

subatomic particles to be found. The link between electricity and the electron was undeniable with the particles appearing to be ripped out of the cathode. Very soon beta radiation was matched to the electron. Radioactive elements were also discovered to release electrons as well. Alpha particles were immensely heavy compared to the electrons – almost 4,000 times heavier – and they were positively charged. These two discoveries would eventually reveal the electromagnetic nature of atoms.

CLASSIFYING STARS

From the earliest days it was clear that not all stars were alike. Some were brighter than others, of course, some appeared to fade away only to return to their original brightness, and some most distinctive stars were known by their red, yellow or blue lights. By the 1860s, the work on spectroscopy pioneered especially by Gustav Kirchhoff offered astronomers a way of studying starlight in great detail. The pattern or spectra of wavelengths in their light showed what elements were present and offered a fingerprint of each star.

The first person to capitalize on that was Father Angelo Secchi, an Italian astronomer (and priest). He started with three classes of star: Class I were blue or white and their spectra showed large amounts of hydrogen. Class II were yellow stars with a lot of metallic elements. Class III were orange and had a wide range of elements present. In 1868, Secchi added Class IV. These were red stars with evidence of carbon. Secchi stellar classification would be steadily refined over the next 50 years and in so doing would reveal the broad structure of the galaxy.

CATALOGUE THE SKY

Star maps live and die by their precision. The history of astronomical discovery can be tracked to the preparation of the upgraded catalogues of the heavens. The Henry Draper Catalogue, largely overseen by Edward Pickering at the Harvard College Observatory, took more than 30 years to produce (from 1886 to 1918). Pickering created an empirical system for measuring star magnitude, or brightness, and he also photographed the spectra of each star observed so it could be analyzed in detail.

Star maps have come on a long way since this 17th-century Dutch map of the constellations.

All this work was carried out by "Pickering's Harem", which was an acceptable term in the 1890s that referred to a team of women mathematicians who computed the data. (Pickering found male assistants inefficient.) Several of the so-called Harvard Computers made immense discoveries, not least Henrietta Swan Leavitt who added the crucial piece to the jigsaw of the expanding Universe, more of which later. Another – Annie Jump Cannon – is credited with being the leading figure in the creation of the Harvard Spectral Classification system which she developed in 1901.

Cannon landed the job of classifying the 220,000 stars that were to appear in the catalogue in 1896. By that time the process was in a mess because the team was unable to agree how to proceed. Antonia Maury wanted to classify stars by temperature and their spectral colours, while Pickering and the more senior computers wanted to extend Secchi's system into 17 classes. Maury refused to do the work, left the project and was replaced by Cannon. She inherited the 17 letters (A to Q) from the earlier prototype version, and slimmed them down to nine: A, B, F, G, K, M and

Annie Jump Cannon.

1900 The Canadian–American Simon Newcomb measures the angle of Earth's axis to the orbital plane around the Sun. It is 23.5° from the vertical.

1901 Ernest Rutherford and Frederick Soddy find that radioactivity makes atoms on one element transmute into atoms of another element. This is a process investigated by alchemists and long thought to be possible. The radioactive emissions change the number of particles in the atomic nucleus, and this alters the elemental nature of the atom.

O, with P and Q retained for peculiarities. The classes were all based on the presence and intensity of emission lines associated with hydrogen, and when she organized these types according to temperature she ended up with a list of O, B, A, F, G, K, M. In keeping with the sexual politics of the time, this was remembered with the mnemonic "Oh Be A Fine Girl, Kiss Me." Later she added the numbers 0 to 9 to indicate temperature, with 0 being the hottest. In 1943, the modern Yerkes classification added luminosity to Cannon's system as a measure of size and brightness. This extra dimension is represented by Roman numerals after the Harvard symbols. So, the Sun is a G2V star, which indicates that it is a yellow dwarf star with a surface temperature of about 5,800K.

Edward Pickering with the Harvard Computers.

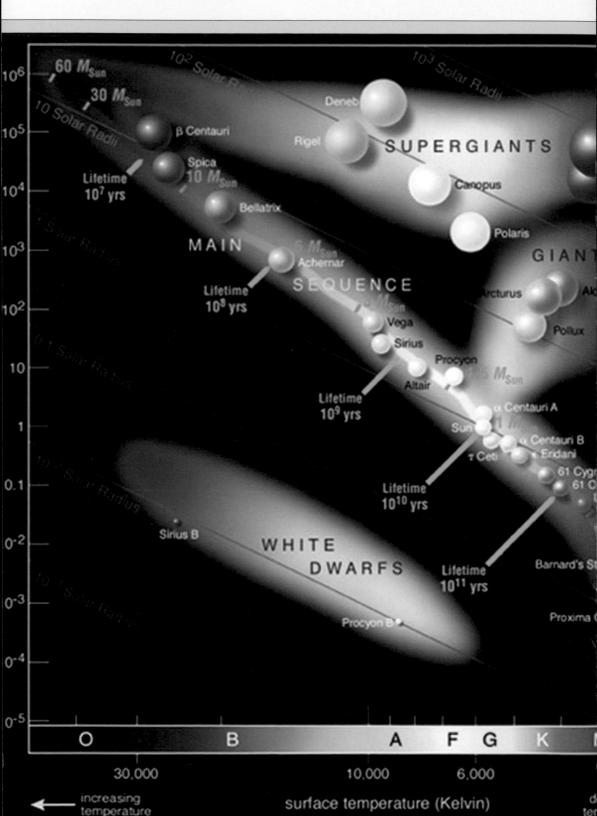

MAIN SEQUENCE

Working independently, the astronomers Ejnar Hertzsprung, a Dane, and Henry Norris Russell, an American, took a different approach to understanding the range of star types. They both plotted the magnitudes (brightness) of stars against their temperatures. The data did not create a random spread, nor did it show a standard deviation with medium sized stars being more common than small and big ones. Instead the graph, published in 1913 and known as the Hertzsprung-Russell, or H-R, diagram, showed most stars, including the Sun, formed a "main sequence" that ran from hot and bright stars (the blue and white ones) to cool and dim (orange and red) ones. Main sequence stars were dwarfs, in relation that is to a cluster of giant and supergiant stars above them. A hot cluster was seen beneath the main sequence formed from hot but faint stars. (In 1914 these were revealed as white dwarfs, the cores of long dead main sequence stars.)

The H-R diagram became a map to guide future investigations into how different stars were formed and where the Universe came from in the first place.

The Hertzsprung-Russell star diagram groups stars according to their size, temperature, and brightness.

1901 The first radio signal is sent across the Atlantic Ocean from Cornwall, England, to Newfoundland, Canada, via technology developed by Guglielmo Marconi.

1902 The Antikythera mechanism, a clockwork device, is recovered from a 2,100-year-old shipwreck in Greece. Later analysis suggests the mechanism is an astronomical instrument.

1903 The first flight of a powered fixed-winged aircraft is made by the Wright Brothers.

1904 J. J. Thomson proposes the plum pudding model of the atom, where negative electrons are suffused through a positively charged medium.

1905 The *annus mirabilis* of Albert Einstein in which he presents papers on the photoelectric effect, Brownian motion, the relationship between energy and matter ($E=mc^2$) and the special theory of relativity.
• The word "genetics" is coined by William Bateson.

1906 Santiago Ramón y Cajal describes the structure of the nerve cell, including the synaptic gap between cells.

1907 The Geiger–Marsden experiment reveals the positively charged atomic nucleus and leads to Rutherford's planetary model of the atom.
• Ivan Pavlov discovers how animals can be conditioned to learn behaviour using a stimulus and reward.

1908 Henrietta Swan Leavitt discovers the link between the time it takes for a Cepheid variable star to grow in brightness and then fade again and its size. This makes this type of star a standard candle, meaning its distance can be calculated from its brightness.

1909 Fritz Haber develops the Haber Process which is an industrial system for producing ammonia from nitrogen in the air. The ammonia is used in artificial fertilizers and explosives.
• Robert Millikan measures the size of the charge of an electron and finds that it is a constant value.

At the turn of the 20th century, physics — and with it the rest of science — was riven with an immense contradiction. But most of the scientific community simply shrugged it off. They believed that physics had solved the mysteries of the Universe; they already had all the answers, and any discrepancies in experiments were simply due to measurement errors. Famously, Albert Einstein did not agree.

In 1905, while working as an office clerk in Bern, Switzerland, Albert Einstein, still only 26, had published four scientific papers. That year became known as his *annus mirabilis*, or "miracle year," because

Young Albert Einstein.

all the discoveries were game-changers, and all were worthy of a Nobel prize. In the end Einstein won the 1921 prize for his description of the photoelectric effect, which helped to raise the veil over the structure of the atom. He also offered the first directly observable evidence of atoms through a phenomenon called Brownian motion. Thirdly he linked mass (m) with energy (E) in his famous equation $E=mc^2$. The fourth paper concerned the "c" in this equation, which denotes the speed of light, and this mysterious speed was the cause of the great contradiction that came to define Einstein's work. His solution to it was the theory of relativity.

1910 Umetaro Suzuki isolates thiamine, also known as vitamin B1, which is the first vitamin molecule to be identified.
• Emil Kraepelin describes Alzheimer's disease as the form of dementia.
• Henri Fabre builds the first seaplane in France.

1911 Heike Kamerlingh Onnes discovers superconductivity in mercury cooled to 4.2 Kelvin.
• Eugen Bleuler gives the first definition of schizophrenia as a distinct condition.
• Charles Wilson invents the cloud chamber, a device for tracking the motion of subatomic particles through space.
• Hafnium is the last element with stable isotopes to be identified.

SUPERCONDUCTIVITY

In 1908, the Dutch physicist Heike Kamerlingh Onnes had used a complex refrigeration technique to cool helium gas into a liquid for the first time in the history of the Universe.

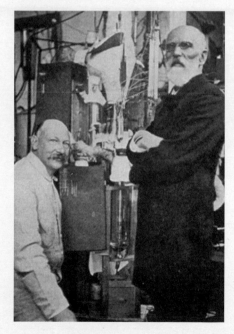

Helium is the least active element of all and thus has the lowest boiling point. Kamerlingh Onnes made the apparatus plunge below the temperature of outer space, reaching 1.5 Kelvin or −272°C, to collect tiny droplets of liquid helium. In 1911, Kamerlingh Onnes was using his deep cold technique to probe the properties of other materials. He froze mercury, a metal that is liquid at room temperature, and plotted the changes in its electrical behaviour as the temperature dropped. He found that at 4.2 K (−269°C) the mercury's resistance vanished. It was conducting electricity with perfect efficiency. This phenomenon was named superconductivity. The deep cold was making quantum effects of atoms appear on the human scale. Kamerlingh Onnes won the 1913 Nobel prize for this discovery, which is predicted to be a crucial aspect of future technology.

Heike Kamerlingh Onnes (seated) at work.

ALBERT EINSTEIN (1879–1955)

Born in Germany but spending most of his youth in Switzerland, Albert Einstein did not thrive at school. He found it boring and was frequently castigated by his teachers. Little did they know that he was already way ahead of them, tinkering with ideas about time and space during his teenage years. As a young adult, Einstein settled on a career as a teacher, but failed to make the grades even then. Something of a Lothario, he was married and a father by 1904 and working in the patent office in Bern to support his family. After the success of his home-grown 1905 papers, Einstein entered academia, eventually settling in Berlin, where he presented his general theory in 1916. Although he was non-observant, Einstein was of Jewish heritage, and in 1933, the Nazi threat forced him to move to the United States. He spent the rest of his career at Princeton University, mostly attempting to link relativity with quantum mechanics. He failed, and that quest continues.

Albert Einstein giving a lecture in Vienna in 1921.

1912 Victor Hess detects exotic charged particles in the high atmosphere, the first evidence of cosmic rays.
• Henrietta Swann Leavitt discovers cepheid variables, stars that change in brightness according to their size. This allows them to be used as "standard candles" or objects the absolute distance of which can be measured by their apparent brightness. This leads to discoveries about the true scale of the galaxy and Universe.
• The parachute is invented.

1913 The Hertzsprung-Russell diagram is used to group stars according to their size, temperature, and brightness.
• Ernest Rutherford, Hans Geiger and Ernest Marsden show that the atom is largely empty space with a positively charged nucleus at its centre.
• Philosopher and mathematician Bertrand Russell reduces mathematics to a set of logical principles.
• The 60-storey Woolworth Building is completed in New York City.

• Niels Bohr publishes his model of the atom with electrons positioned in orbits around the nucleus.

1914 The Panama Canal opens, connecting the Atlantic and Pacific Oceans.
• Karl von Frisch demonstrates that honey bees have colour vision.

1915 Einstein's theory of general relativity explains how space and time can be warped.
• Proxima Centauri, the closest star to the Solar System, is discovered as the third member of the Alpha Centauri system.
• Pyrex heat-resistant glass is invented for use in home and in industry.
• Alfred Wegener proposes that Earth's land was once a single continent called Pangaea as part of his theory of continental drift.
• Junkers make an all-metal aircraft with unsupported wings, very advanced for its time.
• Chlorine gas is used as the first chemical weapon by the German Army.
• The millionth Model-T Ford is manufactured.

1916 Karl Schwarzschild uses Einstein's theory to predict the existence of black holes.

The contradiction Einstein took on was a mismatch between the work of the two great physicists of previous generations: the laws of motion of Isaac Newton and the laws of electromagnetism as defined by James Clerk Maxwell. Newton described nature in terms of matter moving like clockwork. Every motion of one object could be measured as relative to another. For example, cars moving side by side at the same speed, 10 m/s, are not moving relative to one another; their relative speed is 0 m/s. However, two cars moving toward each other at 10 m/s have a relative speed of 20 m/s.

Maxwell's theories concerned a very different part of nature: electric and magnetic fields. He explained that light was an oscillation through these fields, and his mathematics showed that the speed of light was always constant.

This put the two great scientists at odds. Maxwell said light shining from the headlights of one car would always reach the approaching one at the constant speed of light. Newton's laws disagreed, they said that the relative speeds of the two cars would alter the measured speed of light beams. The problem was that no one was able to measure changes in the relative speed of light. Light was always moving at the same speed, even when its source in motion was toward or away from the observer. Einstein wanted to know why.

Portrait of Isaac Newton.

1917 Mark I Tank deployed by British Army, heralding a new era of mechanized warfare.
• Gilbert Lewis proposes the electron shell model of chemical bonding, where atoms share their electrons to form stable connections.
• Albert Einstein publishes what becomes known as the general theory of relativity.

1917 Ernest Rutherford shows that atoms can transform from one element into another through radioactive emissions.

1918 Edwin Howard Armstrong invents the superheterodyne circuit, an essential part of analogue radio receivers.
• "Spanish flu" is first reported in Kansas. Within two years the disease kills an estimated 500 million people around the world.

• First toaster that cooks bread on both sides at the same time is released.

1919 John Alcock and Arthur Whitten Brown, two British pilots, make the first non-stop transatlantic flight from Newfoundland to Ireland in a Vickers Vimy bomber.
• Lee De Forest patents a technique for recording sound on motion picture film.
• Arthur Eddington observes the curvature of light around the Sun in accordance with Einstein's theory of relativity.

1920 Modern research proposes that the HIV pandemic begins in what is now Kinshasa in the Democratic Republic of Congo.
• First two-way radio transmission is made.

THOUGHT EXPERIMENTS

At the age of 16, Einstein asked himself a question: "What would I see if I was sitting on a beam of light?" In the Newtonian context, Einstein would be travelling at the speed of light. Light coming from in front would reach his eyes at twice the speed of light. When looking back, Einstein would see nothing at all. Even though light from behind was travelling at the speed of light, it could never catch up. This was perhaps Einstein's first thought experiment, and he used more to flesh out what became the theory of relativity.

One thought experiment sees a man, Bob, waiting for a train. Night falls and a light on the platform is turned on. Bob measures the speed of the light beam as it moves along the platform. A train travelling at close to the speed of light passes by. Another observer, Belle, on the train measures the speed of the platform light. In a Newtonian universe, the beam should be steadily overtaking the train, but

at a much slower relative speed compared to the platform. However, Belle gets the same value for the speed of light as Bob.

Einstein had to question our most basic understanding of nature to explain why. Speed is a measure of distance against time. Therefore light speed is a constant because time flows at different rates, and the dimensions of space shrink and expand. Objects travelling faster through space move more slowly through time. Clocks on the station and on the train are not moving at the same rate. On the platform, Bob's watch is ticking away as normal, but he sees the clock on the train moving very slowly.

Belle does not notice any slowing of time. The natural oscillations that we use to keep time – the swing of a pendulum, vibration of a quartz crystal or energy fluctuations in atom – are phenomena that obey universal laws. According to relativity, laws remain unchanged within the reference frame.

GENERAL THEORY OF RELATIVITY

After struggling with the many facets of the theory for more than a decade, Einstein finally published the general theory of relativity in 1916. This merged his earlier ideas about the compression and dilation of space and time with energy. Energy — and that includes mass which is simply a form of highly condensed energy — curves space and time. We can visualize this in three dimensions as a flat rubber sheet bulging downward when a heavy weight is placed on it. A straight line that runs across the flat sheet has been curved by the warping effects of the mass. The curve of space makes smaller objects move toward the larger ones, which is analogous to the workings of gravity. According to Einstein, the pull of gravity is due to the warping of space and time by energy.

James Clerk Maxwell.

When an object moves, its energy increases, and in the very simplest sense it gets heavier — or at least it starts to behave as if it is heavier. It takes more energy to shift this "heavier" object to ever greater speeds. Once it gets close to the speed of light, our object has a near infinite mass — or it behaves as if it does. To push it to the actual speed of light would require infinite energy — obviously not something that is possible. In addition, its passage through time has slowed as it accelerates. Now at close to light speed, it is moving through time so slowly that we would have to watch for an infinite amount of time to see it attain light speed. As a result, Einstein declared that the speed of light is the speed limit through space — and that nothing with mass can go that fast.

177

PROOF

Einstein's ideas were met with amazement and indignation. It all sounded too wacky for most. Then in 1919, the English astronomer Arthur Eddington showed that relativity's way of describing the Universe was true. He was observing a full solar eclipse, a rare occasion when the glare of the Sun is blocked out to reveal the more diffuse layers around it. Eddington saw that stars located in line with the Sun's edge appeared out of place. The great mass of the star was warping

Headlights give out beams at the speed of light no matter how fast the cars and trucks are moving.

space around it, and the light travelling through curved away from its straight path — and thus appeared in a different place in the sky.

Since then, Einstein's ideas have been proved right countless times more

We use his theory to understand black holes and to make GPS satnavs work. In 2016, waves in space–time as predicted by relativity were discovered for the first time. They now offer a new way of imaging the Universe that will show us what light and other radiation cannot.

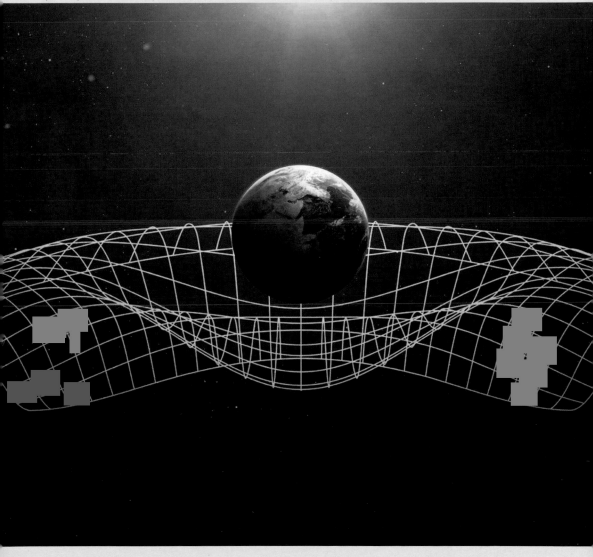

Space warps around massive objects.

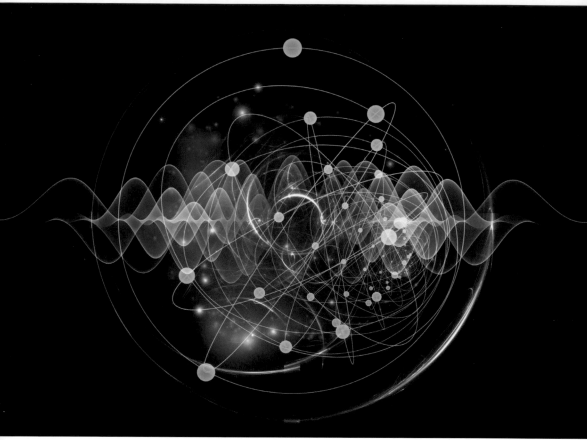

The mathematical tool of waveforms helped quantum physicists to explore the properties of particles.

Once it was clear that atoms were not the smallest thing in the Universe and instead were made from smaller particles, the question was how these were arranged. J. J. Thomson had shown that atoms contained negatively charged electrons but overall were electrically neutral. In the absence of any other evidence. Thomson suggested that atoms were made from a positively charged medium dotted with negatively charged electrons, like plums in an English desert.

The "Plum Pudding Model" was soon superseded by Rutherford's discovery of the atomic nucleus in 1913. Rutherford's new synthesis of the atom was an orbital system where the negative electrons moved around this tiny positive core, where most of the atom's mass was concentrated. It would be another four years before Rutherford could isolate the source of the the positive charge — the proton. Even then that particle did not supply enough mass to account for atomic weights. The obvious solution was the presence of an electrically neutral particle with a very similar mass to the proton – soon termed the neutron. This was indeed the case but it took until the mid-1930s to isolate it.

ATOMIC NUMBER

Even prior to these subatomic revelations, the nature of the atom was being deduced. In 1913 Englishman Henry Moseley, a protégé of Rutherford, found that he could correlate the wavelength of X-rays given out by atoms to the positive charge of their nucleus. He worked mostly with metals, and the increase in X-ray wavelength was linear as he worked his way up from lighter elements to heavier ones. He used this fact to give each element an atomic number to represent the nuclear charge, starting with 1 for hydrogen, 2 for helium, and so on through the elements.

Moseley's atomic number proved a very effective way of organizing the periodic table, because unknown to him he had revealed a central property of the atom. The working theory of the time was that heavier elements had atoms containing more protons (still to be proven, remember), and so the greater charge of the larger nuclei corresponded

1920 Frederic Clements divides up the world's habitats into large-scale groups termed biomes. Biomes cover areas with broadly similar climates and present particular challenges to wildlife.

1921 Psychologist Hermann Rorschach presents his now famous inkblot tests as a way of probing the subconscious of his patients.

1922 The last Barbary lion, the North African subspecies of lion, is shot in the mountains of Morocco.

1923 Karl von Frisch describes the dance of the honey bee that the insects use to communicate the location and size of pollen and nectar sources to other members of the the hive.

1924 The structure of the cell membrane is discovered.
• The concept of food webs is proposed by Charles Elton and Alister Hardy.
• Louis de Broglie proposes that particles are better understood as matter waves, or waveforms.

1925 Edwin Hubble finds objects beyond the edge of the Milky Way, showing that ours is not the only galaxy in the Universe.
• Wolfgang Pauli outlines the exclusion principle, one of the founding ideas in quantum physics.
• John Logie Baird invents the form of television which is used in the first TV broadcasts.
• Satyendra Nath Bose collaborates with Albert Einstein to describe a set of particles that carry forces between massive particles. These particles are named bosons in his honour and include photons and gluons.
• Cecilia Payne shows that stars are made almost entirely of hydrogen and helium.

Henry Moseley showed that each element has a unique atomic number.

to them having more protons. The negative charge of an electron balanced that of a proton, and so the number of electrons in the atom was always equal to that of protons. Therefore, the atomic number was really the number of an atom's protons. Moseley did not live to see this simple and exquisite atomic model confirmed because he died in 1915, shot by a sniper during World War I, before its final pieces could be slotted into place.

QUANTIZED ATOM

Another student of Rutherford and colleague of Moseley at Cambridge was the Dane Niels Bohr. He wanted to understand how the motion and position of electrons related to the way atoms received and released energy in the form of electromagnetic radiation. Rutherford's idea of an atom had electrons whizzing around the nucleus like planets around a tiny star. Bohr tried out a model where the electron's energy was proportional to the orbital frequency of the electron – which was derived from its speed and distance from the nucleus. The speed and position of an electron would be often changing as

it absorbed energy and emitted it again as light or whatever. So Bohr proposed that the constant that linked an electron's energy and its orbit was some fraction of Planck's constant, which forms a similar link between the energy of a light wave and its frequency.

Bohr then applied this system using the known frequencies of different elements' spectra – the unique sets of light they absorb and emit – and he found that he could only make the maths work if electrons occupied certain positions around the nucleus. It was impossible for the electron to take a position halfway

Niels Bohr (left) with Max Planck.

1926 Robert Goddard invents the liquid-fuelled rocket.

1927 Werner Heisenberg sets out the uncertainty principle which explains the limits of how accurately quantum states can be measured.

• George Thomson performs a double slit experiment with a source of electrons. The slits create an interference pattern similar to that produced by a light wave. This shows that electrons are indeed particles and waves at the same time.

1928 English physicist Paul Dirac creates a mathematical description of the electron. This equation is extended to all electron-like particles, and also shows the existence of antimatter.

between or anywhere else but these locations, which are now termed orbitals.

That was because radiation of a specific frequency contains an specific quantum of energy. To move an electron from a low orbital to a higher one could only happen if the precise amount of energy was delivered by the precise frequency of light. Then it would make a "quantum leap" to its new orbital. This is why atoms have specific absorption spectra. They can only take in the energy provided by a very particular set of frequencies.

The energy in the light allows the electron to move slightly further away from the eternal attraction of the positive nucleus, and leap further out to a new orbital. When the electron leaps back to a lower position, it will give away the extra energy as a single packet of energy, a photon.

ELECTRON SHELLS

The orbitals of an atom are arranged in distinct layers or shells, with each shell having space for more electrons than the one further in. The number of spaces in shell n is always $2n^2$. The first shell has two spaces (2 x 1 x 1), the second one has 8 (2 x 2 x 2), the third 18, etc. The largest atoms have seven electron shells. Each element has a unique number of electrons in its atoms and a unique electron configuration. This configuration is what is being shown in the periodic table. The row or period of an element indicates how many shells its atoms have. The group indicates how many electrons are held in the outermost shell. These outer electrons get involved with bonding to other atoms during chemical reactions and their number corresponds directly to the older and now defunct concept of valence.

Electrons can be understood as areas of charge that fill spaces around the nucleus.

183

WAVEFORMS

Light was a wave for sure, but Einstein had also shown that its quantum nature was from photon particles. In 1923, the Frenchman Louis de Broglie proposed that this wave-particle duality was not exclusive to radiation. All subatomic particles that operated at the tiny quantum scales could be described as a waveform. They did not radiate across the Universe at the speed of light, but de Broglie encapsulated every property of the particle, its speed, spin, charge and kinetic energy, in the mathematics of oscillation.

The mathematical tool of waveforms allowed quantum physicists to begin to explore the properties of particles. In 1925, Wolfgang Pauli discovered the exclusionary principle. Every atom has a finite number of quantum states that an electron can occupy, and the principle says that only one electron can be in each state at a time. The principle was later extended to all particles that make up regular matter.

In 1927, George Thomson, son of J. J., turned this mathematical waveform idea into physical fact. He repeated the Young double slit experiment, which had shown light to be a wave, but used a beam of electrons instead. As particles, electrons would go through one or other of the slits. This would show up and two clear bands on the detector. However, Thomson saw that the electrons made an interference pattern just like a wave of light – and so the electron was a particle and a wave at the some time.

UNCERTAINTY

In 1927 Werner Heisenberg, a young researcher at the new Niels Bohr's Institute of Theoretical Physics in Copenhagen, along with a fellow quantum physicist, Max Born, showed that the wave-nature of matter was

Every founding figure in quantum physics attended the Solvay Conference in 1927. Werner Heisenberg is third from the right in the back row. Max Born is seated in front of him.

inherently probabilistic. The properties of quantum objects were not fixed but had a particular chance of being in one state or another, which was all wrapped up within the mathematics. Only when you measured or observed the state would you know which one it was. The probability was now replaced by certainty and the quantum waveform had collapsed. However, the waveform linked properties together, and so by observing one property, such as the location of an electron, it became impossible to ever know its momentum (the speed and direction). This became known as the Uncertainty Principle and it creates some of the weirdness of quantum

1929 Karl Lohmann, Cyrus Fiske and Yellapragada Subbarow all find adenosine triphosphate (ATP) in cells. This molecule is used to store and transfer energy around a cell.

physics: particles can exist in multiple states all at once, a phenomenon called superposition, and it is impossible to link cause and effect, because at the quantum level it is all down to chance — it could go either way and we can never know until it happens. One of the central tenets of physics is that the present is the result of events in the past, and the events taking place now will cause the future. Quantum physics does not agree.

CHEMICAL ENERGY

In 1929 several biochemists discovered the chemical adenosine triphosphate in cells — no matter what cell they looked in, there it was. To save everyone time, the compound name was shortened to ATP, and it was later shown to be the chemical that holds the energy that keeps us all alive. The cell's mitochondria extract energy from food using respiration, and that free energy needs to be captured, stored and transferred. The energy from respiration is used to attach a third phosphate to ADP (adenosine diphosphate, the "D" showing it has one less phosphate group than ATP) charging up the molecule into ATP. When needed, the ATP ejects a phosphate, releasing a packet of useful energy to the metabolic process taking place in the cell. The resulting ADP returns for recharging. One molecule of glucose can charge up 30 ATPs.

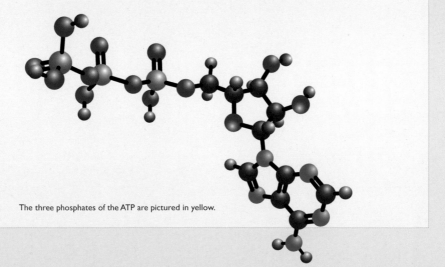

The three phosphates of the ATP are pictured in yellow.

The Milky Way.

As the 1920s progressed, astronomers felt they were closing in on a definitive answer to the size and extent of the Milky Way. In 1904 the Dutch astronomer Jacobus Kapteyn revealed that from the point of view of Earth, stars appeared to stream in either one direction or another – never radiating off in random directions. Twenty years later Bertil Linbland, a Swedish astronomer, deduced this meant that our Solar System was part of a vast rotating disc. Stars further out from the centre appeared to lag behind, while those further in than us streamed ahead of our motion around the central point. This discovery resulted in the most extensive survey of the Milky Way to date. Linbland also discovered that the middle of the Milky Way, the region of our sky that is packed most densely with stars, was a dense bulge rising out of the disc on both sides.

This disc structure of the galaxy reminded astronomers of the spiral nebulae that could be seen with the most powerful of telescopes. Were these fuzzy structures within our own galaxy or around the edge, or were they galaxies in their own right, located far beyond our own across a great gulf of emptiness? Quite a question.

Vesto Slipher, the chief astronomer at Lowell Observatory, Arizona.

1929 Edwin Hubble discovers that the Universe is getting larger. He is able to show that all galaxies are moving away from each other, and the ones that are the furthest away move the fastest.

• The term "homeostasis" is coined by Walter Bradford Cannon to describe the self-limiting processes that maintain stable conditions in a living body.

• The Van de Graaff generator is invented as an upgraded version of a friction-powered electrostatic generator. It is capable of generating charges of millions of volts.

• The first circumnavigation of Earth by air is achieved by the German airship Graf Zeppelin in 1929. It takes 21 days to complete the journey.

1930 Linus Pauling gives the best description yet of how chemical bonds form between atoms due to the configuration of electrons in their outer orbitals or shells.

• Pluto is discovered by Clyde Tombaugh at the Lowell Observatory and is designated as the ninth planet. (Today Pluto is classed as a dwarf planet.)

DISCOVERING PLUTO

The Lowell Observatory in Arizona, set up originally to find evidence of Martians, shifted its gaze at the start of the 20th century. It began to survey the ecliptic – the strip of sky occupied by the planets – for Planet X in 1906. Ten years later Percival Lowell, the founder, died. A dispute with his widow over funds put a stop to this and several other projects until 1929. Then Clyde Tombaugh, a 23-year-old astronomer, was given the job of looking for a new planet. Over the course of a year, he took pictures of the ecliptic every two weeks, and compared them to see if anything moved out of position against the background of stars. In 1930 he spotted an object that appeared to be following a planetary orbit. Following a public consultation, it was named Pluto thanks to the contribution of a Venetia Burney, an 11-year-old English girl. Pluto was initially judged to be the size of Earth but 76 years later it was found to be barely a fifth of the mass of our Moon. Astronomers demoted tiny Pluto to the status of dwarf planet. Planet X, if it exists, is now predicted to orbit three or four times further away from the Sun than Neptune.

Clyde Tombaugh with the telescope used to spot Pluto.

REDSHIFTS

Vesto Slipher, the chief astronomer at the Lowell Observatory, had mapped the redshift of objects across the sky, including big nebulae like Andromeda (which is six times wider than the Moon, just too faint to see.) Redshift is the astronomical version of the Doppler effect. The Doppler Effect, named after Christian Doppler who discovered it in 1842, stretches or compresses waves according to the relative motion of the source and observer. Best experienced with sound, the wavelengths of the siren of an approaching ambulance are compressed and so they sound higher pitched. When the vehicle races past and away from you,

Henrietta Swan Leavitt.

STANDARD CANDLES

The answer was supplied by the work of Henrietta Swan Leavitt who had worked for many years away from the limelight as one the Harvard Computers. These mathematically-minded women crunched the reams of data being collected to produce the Henry Draper Catalogue, which represented a step-change in star mapping. One of Leavitt's jobs was to quantify the brightness of stars, and she took an interest in Cepheid variables, a particular kind of star that grew brighter then faded in a periodic fashion. Leavitt discovered that time between peaks in brightness of a Cepheid – they were first seen in the constellation Cephus and were named accordingly – was correlated to the size of the star. Bigger stars took longer to brighten up and fade away.

This provided a crucial observable link between actual size of a star and its distance. A large, bright star may appear very dim because it is so far away. A small dim star is still very bright when close to our Solar System. There was no way of knowing which was which simply by looking. The Cepheid variable, however, was a "standard candle." You could figure out how far away a star was just by looking.

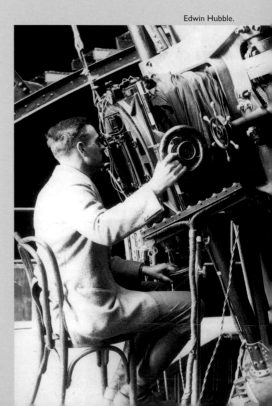

Edwin Hubble.

its sound waves are stretched instead, and the pitch of the siren drops. The same thing happens to the wavelengths of its light from a star, specifically the absorption spectra of its atmosphere.

Slipher found that Andromeda's light was more blue than expected – its light waves had been compressed. That meant Andromeda was moving toward us. However, Slipher found that most of the hundreds of other nebulae he surveyed were redshifted. Their light was being stretched to longer, redder wavelengths because they were moving away from us. The degree of their redshift was not uniform, however. What did that signify?

1932 English physicist James Chadwick discovers the neutron, an electrically neutral subatomic particle.
• The positron, the antimatter version of the electron, is discovered. Both particles are identical but the position is positively charged. Positrons are outnumbered by electrons and are very short-lived. When it meets an electron, the two particles annihilate each other.
• Fritz Zwicky proposes the existence of dark matter.

1933 Subrahmanyan Chandrasekhar calculates the smallest size that a star must be to become a supernova (around 1.4 times the mass of the Sun).
• Walter Baade and Fritz Zwicky propose that most supernovae result in an ultra dense body called a neutron star.

1934 Pavel Cherenkov explains that radioactive particles can break the speed of light in certain media. This creates a tell-tale flash of radiation that can be used to identify the presence of subatomic particles.

CHERENKOV RADIATION

Light's top speed is through a vacuum. Nothing can go faster than that. However, light passing through a transparent medium is forced to slow slightly. This effect is the cause of the phenomenon of refraction as waves are deflected slightly and they speed up and slow down at the interface between two media. For example, the speed of light in water is just three-quarters of what it is in a vacuum. In 1934, the Russian Pavel Cherenkov figured out that radioactive particles could be blasted out so powerfully that they moved faster through a particular medium. In the case of water, such as the coolant around nuclear fuel rods, this effect created an eerie blue glow. The high-speed particles rip through the surrounding atoms transferring their energy that is given back as a light. Tiny blinks of Cherenkov radiation became a way of seeing tiny particles and is used to this day as evidence of high-speed collisions inside high-tech particle detectors.

Cherenkov radiation in a research reactor.

UNIVERSAL EXPANSION

In 1925, an American astronomer Edwin Hubble found Cepheid variables in the Andromeda and Triangulum nebulae. He was working with the Hooker Telescope, located on Mount Wilson in California and the largest observational instrument of its day. The variable stars demonstrated to Hubble that they were far beyond the edge of Milky Way, and that meant that the nebulae were not cloudy star clusters but immense galaxies, Andromeda galaxy being even bigger than the Milky Way. Four years later, in 1929, after extensive research into the location and distance of other galaxies, Hubble was able to link their distances with their redshifts. The bodies that were further way had higher redshifts, and that indicated they were moving away faster than those objects nearby.

Galaxies were not just moving away from us, they were all moving away from each other, too. Hubble's conclusion resonates to this day: the Universe is getting bigger all the time. And it is not simply spreading out into previously empty space; the empty space is expanding as well!

A BIG BANG

If the Universe is expanding as it gets older, it must have been smaller when it was younger. If we could rewind history then eventually we would arrive at a time when the Universe took up a single point in space. Even before Hubble's observational evidence of universal expansion, a Belgian priest, Abbé Georges Lemaître, had used Einstein's mathematics to come to the same conclusion. The Universe could never be still, so was either expanding or contracting. In 1931, Lemaître suggested that the Universe began as a "primeval atom" which exploded outward giving rise to all the other atoms in the Universe we see today. There was no evidence for this idea, but it had a certain creative spark to it.

The "Big Bang" as it has become known has been the enduring theory of the Universe ever since. The events following the Big Bang were first sketched out in the late 1940s by Ralph Alpher and George Gamow. They added the name of their friend Hans Bethe, who was famous for unpicking the process of nuclear fusion that powered the Sun and other stars, to create the Alpher-Bethe-

Abbé Georges Lemaître.

Gamow paper, a pun that linked the ABC of the Greek alphabet and the primitive Universe. The paper proposed how the Big Bang created a seething hot universe filled with hydrogen atoms, which has expanded and cooled ever since into the stars and galaxies we see today.

Despite its great complexity to the lay person, this was a rather simplistic description of the Big Bang. Nevertheless, every generation of physicists and astronomers has tested the idea and

1935 Nylon is invented.

1936 The mu meson, or muon, is seen for the first time in collisions of cosmic rays in the high atmosphere.

• The Hoover Dam is completed across the Colorado River between Nevada and Arizona.

1937 Element number 43 is discovered for the first time in residue made by early particle accelerators. The element is too unstable to exist in nature, but it is named technetium as the first artificially produced element.

• Hans Krebs discovers the citric acid cycle which is the sequence of chemical reactions involved in respiration, the process that releases energy from glucose to power every living cell.

• The Hindenburg airship crashes in New Jersey bringing the era of long-distance airship travel to an end.

1938 Italian nuclear physicist Enrico Fermi sets off the first nuclear fission chain reaction, where a uranium atom splits into two smaller atoms of different elements. The split releases neutrons, which hit other uranium atoms making them split and continue the reaction.

1939 Superfluidity, where a liquid moves without friction or surface tension, is discovered by cooling helium to very low temperatures.

1939 Francium is the last element to be discovered in nature, rather than synthesized in the lab.

• Hans Bethe explains how stars produce energy through the nuclear fusion of hydrogen.

• The Heinkel He 178 is built by the German air force. It is the first jet-powered aircraft.

provided compelling evidence for an ever-more detailed description of the process. Still today, the biggest science experiments on Earth are looking for further details about the earliest moments of the Universe to help understand its present and predict its future.

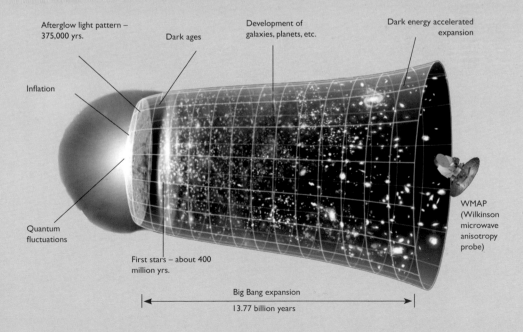

Afterglow light pattern – 375,000 yrs.

Dark ages

Development of galaxies, planets, etc.

Dark energy accelerated expansion

Inflation

Quantum fluctuations

First stars – about 400 million yrs.

WMAP (Wilkinson microwave anisotropy probe)

Big Bang expansion

13.77 billion years

The 1940s saw the realization of a long-held medical goal: the magic bullet. The magic bullet is a single drug that can tackle the causes of infection without damaging healthy body tissue. A suitable candidate for this medical weapon, penicillin, had been known for almost two decades, but by 1944 a method of its mass production had been perfected. This breakthrough came not a moment too soon for the millions who would have otherwise succumbed to infected wounds during the Second World War. The new germ-killing drug, an antibiotic, would also be crucial during the reconstruction, producing the healthiest generation in history, boosting life expectancy, and helping to quadruple the human population in little more than 70 years.

SPREADING INFECTION

Penicillin was the culmination of a two-pronged attack from the combined forces of medical science and biology. In 1847 Ignaz Semmelweis, a Hungarian physician, had wondered why so many new mothers suffered from so-called "childbed fever," a sudden and violent disease that killed them a few days after giving birth. His suspicion was that the doctors who attended to the women during childbirth were spreading the disease – although he did not understand how. At the time, childbirth had become increasingly medicalized with the age-old services of midwives being sidelined. Doctors, invariably male, would take charge of births to ensure a safe delivery with several novel interventions. While these treatments were often based in sound medical evidence, the doctors seldom washed their hands between patients. Semmelweis was able to show that simply disinfecting hands and equipment drastically cut the incidence of fever.

In 1854, John Snow, a founding figure in epidemiology – the science of public health – famously linked an epidemic of cholera in London with the tainted water supplied from a single public pump in Soho – but again could not explain what made the water so deadly. Shortly after, Florence Nightingale used meticulous statistical analysis to prove once and for all the well-

Ignaz Semmelweis linked hygiene to health.

1940 Astatine is obtained by bombarding bismuth with helium nuclei.

1941 Plutonium is prepared by the bombardment of uranium with heavy hydrogen isotopes.

1942 Werner von Braun builds V2 rocket bombs which make the first suborbital space flights.
• Magnetic recording tapes are invented.

1944 Quinine, a herbal antimalarial drug, is synthesized.

1945 Atomic bombs are dropped on Hiroshima and Nagasaki in Japan.

1946 Fred Hoyle and others describe stellar nucleosynthesis, in which all elements heavier than helium are created inside stars.
• Ballpoint pens, invented by Hungarian Laszlo Biro, sell by the million.
• ENIAC is built in the USA, the first general purpose programmable digital computer.

1947 Chuck Yeager breaks the sound barrier with the Bell X-1 rocket plane.
• The polaroid instant camera is introduced.
• The transistor is invented.

1948 The field theory of quantum electrodynamics is developed by Richard Feynman and others.
The Alpher-Bethe-Gamow paper proposes how the Big Bang resulted in the production of the Universe's atoms.

1949 Albert II, a rhesus macaque, is launched aboard a V-2 rocket commandeered and modified by the United States. He flies to an altitude of 134 km (83 miles) and becomes the first mammal in space. Sadly he dies on landing due to a parachute failure.

founded intuition that hygiene and good health were strongly linked. (Nightingale is famous for being a pioneer in modern nursing, but her ability to use medical data to prove her point was her more unsung ability.)

The biologists soon came up with the missing piece of the puzzle. In 1861 Louis Pasteur proposed his germ theory of disease, namely that tiny organisms seen only through microscopes were responsible for infections. This work was later compounded in the 1880s by Robert Koch, one of the first microbiologists, who provided a clear link between particular diseases and microorganisms.

This double decade of discovery did much to help prevent future infections, with the rise of antiseptic practices in surgery and beyond. However, once a patient had an infection, there was little to be done but wait for the immune system to fight it off — if it could.

Louis Pasteur proposed the germ theory of disease in the 1860s.

ACCIDENTAL BREAKTHROUGH

An accident in the summer of 1928 was the breakthrough that changed all that. Alexander Fleming, a bacteriologist working in London, returned to his lab after a holiday in his native Scotland to find that fungus had grown over his equipment. He noticed that the fungus had affected the bacteria in the sample dishes he had been studying. No bacterial cultures grew around the fungal invaders, creating distinct germ-free rings. Fleming realized that the fungus, a species of *Penicillium*, similar to the ones used to make blue cheese, was releasing a chemical that killed bacteria. Fleming isolated the substance, which he named penicillin, and did some rudimentary testing on laboratory mice. It proved harmless to non bacterial cells. (Penicillin acts by preventing bacteria from controlling the flow of water into their single-celled bodies. That influx makes them simply burst open.) Fleming reported that he had treated a colleague's eye infection with his fungus-based medicine.

Fleming spent the 1930s trying to find a chemist to analyze penicillin with a view

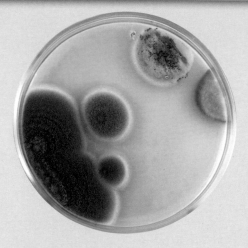

Colonies of *Penicillium* on an agar Petri dish.

to making it on a large scale. It was beyond the scope of all he approached, however. But in 1940 the challenge was taken up by Howard Florey and Ernst Chain, the first a pharmacologist, the second a biochemist working in Oxford. The pair were funded by the British and American governments to seek out a method of purifying the compound in large quantities. The first penicillin plant came on stream in the months after the Pearl Harbor attack, and by D-Day three and half years later, there was enough of this new drug — an antibiotic — to treat all the wounded in the Allied armies.

Fleming, Florey and Chain won the Nobel Prize for this achievement in 1945.

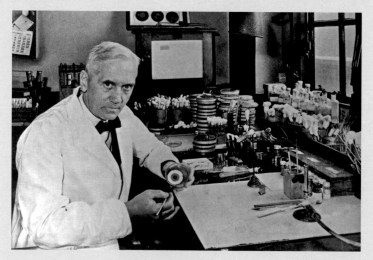

Alexander Fleming discovered penicillin in 1928.

SUPERBUGS

That same year, Fleming made an ominous prediction. He reported that since a low dose of penicillin produced a bacterial population more resistant to its effects, using the antibiotic in the wrong way would lead to the evolution of superbugs, which would be impervious to antibiotics. As Fleming put it: "In such cases the thoughtless person playing with penicillin is morally responsible for the death of the man who finally succumbs to infection with the penicillin-resistant organism. I hope this evil can be averted."

Penicillin saved the lives of many soldiers in World War II, who would have died from infected wounds.

The modern world is a whirlwind of constant change, impacted on all fronts by a near constant stream of new technologies. The computer revolution of the 1970s and 80s gave way to the communications revolution of the 90s and 2000s, and we now stand on the verge of the robotics revolution, so we are told, and from there will be a short step to artificial intelligence. However, in some respects nothing new has been invented since 1948 with the arrival of the transistor.

FAST SWITCH

The transistor is a switch that turns a current on and off. Instead of using a mechanical part as the electromagnetic switch that preceded it, the transistor has no moving parts and can switch on and off much faster. Today's versions can do this 600 billion times a second, although the early version built by William Shockley, John Bardeen, and Walter Brattain,

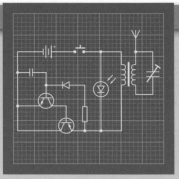

A circuit diagram including transistors.

The very first transistor.

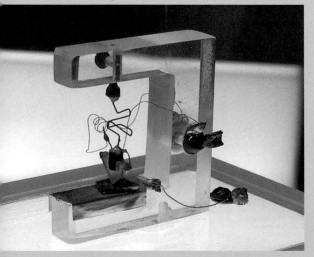

working at the Bell Laboratories in New Jersey, was much slower.

Transistors were designed as the central components of electronic devices. Electronics is a technology that uses the flow of electricity to control a device, and the earliest versions used thermionic valves. Early digital computers developed at the tail end of World War II used them to process instructions inputted as binary codes that used only two numbers, or digits, the 1 and the 0 – hence the term digital.

1950 The Dutch astronomer Jan Oort suggests that long-period comets, which take many centuries to orbit the Sun, originate in a sphere of primeval material that surrounds the edge of the Solar System. This region is now named the Oort Cloud.

1951 The Mark I computer is delivered to the University of Manchester It is the first general-purpose electronic computer that is commercially available.

1952 The process by which the action potential, an electrical signal that travels along nerves, is explained after many years of research.

1953 The Miller–Urey experiment investigates how a mixture of simple chemical changes are repeated inputs of energy. The idea is to emulate the primordial soup, and the experiment shows that complex biochemicals similar to the ones found in living things can form by chance alone.

THE CHEMISTRY OF THE PRIMORDIAL SOUP

Harold Urey was one of the prime movers in the Manhattan Project, developing gas diffusion techniques for enriching nuclear explosives. After the war he was behind an iconic experiment that tested the chemistry of Earth's early atmosphere and oceans, and the results were surprising. He took his cue from the chemistry of other planets and suggested that Earth's early atmosphere had no oxygen but was filled with water vapour, methane, ammonia, and carbon dioxide. In 1953 Urey teamed up with a graduate student called Stanley Miller to make a tiny recreation of Earth's early chemistry, or primordial soup. This chemical mixture was held inside a round vessel, with a loop of tubing to allow gases to circulate. It was then heated, cooled, stirred, and electrified at various points to mimic the conditions of the early Earth. Within a day the initially clear mixture had gone pink. After a week more than 10 per cent of the carbon atoms had formed into complex organic compounds, such as amino acids, the building blocks of proteins. There are about 20 amino acids

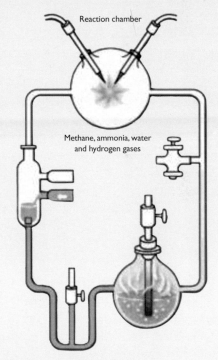

The Miller–Urey "lollipop" apparatus.

used in nature, and the Miller–Urey experiment made 11 of them. The experiment has been re-run many times with different proportions of material and proves at the very least that the component parts of complex biochemistry can arise spontaneously from non-living processes.

DIGITAL TECH

Thermionic valves are high-tech vacuum tubes that could switch a current on or off through the application of a smaller current to a part of the device. Transistors work in basically the same way, but use a highly modified crystal of pure silicon. Pure silicon crystals came to the fore during the development of radar. It was found that pure silicon was the best at amplifying the weak radar signals that echoed back. Silicon is a semimetal element, which means it has properties of both metals and non-metals. Pure silicon

A thermionic valve.

looks rather shiny and metallic and it conducts heat quite well but is not very good at conducting electricity.

SEMICONDUCTORS

A silicon atom has four electrons in the outermost orbitals, which makes the outer shell half full. This means the atoms are unlikely to give any free electrons or accept extra ones, and as a result it takes a lot of energy to force electrons away from the atom so a crystal can conduct a current. This is of no use in electronics. Instead silicon is converted into a semiconductor, a material that can act as a conductor or an insulator in particular conditions. To create a semiconductor, a silicon crystal is doped with tiny amounts of other atoms. In the simplest example, phosphorus is added to create a crystal with a few spare electrons. That makes a substance called an N semiconductor (for negative charge). Adding boron creates a crystal with a lack of electrons. Areas of the crystals that lack an electron are known as hypothetical positively charged entities known as "holes", and they are able to flow around just like electrons, albeit in the opposite direction. A substance with an excess of positive holes is a P semiconductor (for positive charge).

A transistor is one type of semiconductor sandwiched either side

Pure silicon.

with two pieces of the opposite kind of material so it is either NPN or PNP. The central section is the base, and either side is the emitter and the collector. The transistor is switched on by applying a small positive electric charge to the base. In the case of an NPN, where the P base has an excess of positive particles, the positive charge repels the holes and attracts any electrons. This creates a channel of negative charge running through the base from the N emitter to the N collector, and a current can flow through it. So switching on an NPN makes current flow.

1953 Francis Crick and James Watson make public the structure of DNA (deoxyribonucleic acid) as a double helix molecule. This complex molecule is made up of several smaller units and is able to carry a four-character code which is the basis of genetic inheritance.

• Masers, an early form of lasers, are invented.

1954 The Castle Bravo H-bomb was tested in the Pacific. It was the first thermonuclear weapon to use solid fusible material and formed the basis of the world's nuclear arsenals today.

1955 Fred Hoyle and Martin Schwarzschild describe how dwarf stars, like the Sun, end their lives as red giants that are filled by the fusion of helium.

1956 Gilbert Plass presents the most thorough prediction to date of global warming through anthropogenic carbon dioxide and begins the study of climate change.

H-BOMB

By the late 1940s, the theory of nucleosynthesis was developed which explained how all "heavy" elements – from lithium to uranium and beyond – were formed by the fusion of smaller atoms. All that began with the fusion of hydrogen and helium atoms in the heart of a star. Common elements like carbon and oxygen were made when a regular star reached the end of its fuel cycle. The truly rare elements, like gold, were only ever made in the cataclysm of a supernova, which was the implosion of the largest of all stars.

A few years after the American fission bombs, powered by the splitting of large atoms of uranium and plutonium, had ended World War II, a new fusion bomb was developed. Even more powerful, the so-called H-bombs released explosive energy by fusing two small hydrogen atoms. The first H-bomb explosion was on a remote Pacific atoll in 1952. Code-named Ivy Mike, it was a thermonuclear device in that it used the heat energy from a small fission bomb to force a fusion bomb to explode, unleashing almost 1,000 times the power of Little Boy, which had obliterated Hiroshima.

The distinctive mushroom cloud from the first H-bomb nuclear test on Enewetak Atoll in the Marshall Islands, November 1952.

LOGIC GATES

Transistors are grouped together to make logic gates. There are a couple of dozen of these and each one processes inputs in the form of 1s and 0s into outputs according to a mathematical operation defined by Boolean algebra. This strange mathematics uses operations that are sometimes similar to addition and multiplication but in other ways very different. Because it uses only 1s and 0s, the answers to Boolean sums can only ever be 1 or 0. For example, 1+1=1. This system was developed by George Boole in 1854 as a way of making logical decisions using nothing but numbers, and it is the system used in every electronic circuit today.

The first consumer technology to use a transistor was a 1952 hearing aid, and the benefits of these tiny devices were soon manifold. Transistor radios, introduced in 1954, were small enough to be easily portable. In 1959, the first US satellite used germanium and silicon transistors. Around this time, the integrated circuit was invented, where an entire electronic circuit made up of many transistors and other components all connected by conducting wires were placed on a single piece, or chip, of silicon. Integrated circuits were made smaller and smaller, becoming known as microchips. William Shockley, one of the transistor's pioneers, had moved to Mountain View near California in the mid 1950s to be nearer to his mother. He set up a business developing semiconductors, and similar businesses grew steadily in the area, resulting in what is now called Silicon Valley.

One of Shockley's employees, Gordon Moore, who would go on to found the Intel chip makers, predicted that the processing power of microchips would double every 18 months (and the size of transistors would halve). So far, Moore's law has been broadly accurate. Today a transistor's base is about 50 atoms wide, and one chip can contain billions in a single circuit. However, the transistor cannot get much smaller, or it will become too narrow to form a barrier against electricity.

George Boole.

1957 *Sputnik I* is the first spacecraft to go into orbit and becomes the first artificial satellite.
• The laser is invented.
1959 Data from *Explorer I*, the United States' first satellite, reveals that Earth is surrounded by a magnetic field that reaches far into space.

1960 Two Russian dogs, Belka and Strelka, become the first animals to orbit Earth and return alive to the surface.
• The American test pilot Joe Kittinger parachutes from a balloon 31 km (19 miles) above Earth's surface in conditions very close to outer space.

LASERS

A laser is a coherent, collimated light. The term coherent means the light waves oscillate in time with each other, while collimated means the waves are parallel, rather than shining out in all directions. Laser light also contains a handful of wavelengths, perhaps just one, instead of a wide spectrum like a natural light source. All this means that laser light can be used to reflect, refract, illuminate and heat in very precise ways.

Before lasers came masers, which stands for Microwave Amplification by Simulated Emission of Radiation. (Microwaves are high-frequency radio waves.) It was made by energizing a "gain medium" – generally a transparent crystal. The addition of energy makes the medium emit light (or another ray), and this reflects off mirrors back into the crystal, stimulating it more and producing more light. The high intensity radiation directed out of the medium creates a beam. The maser was in 1953, and by 1957 a version using visible light was developed. In keeping with the acronym naming, this device should have been a Light Oscillation by Stimulated Emission of Radiation, or "loser". Instead "maser" was simply changed to "laser".

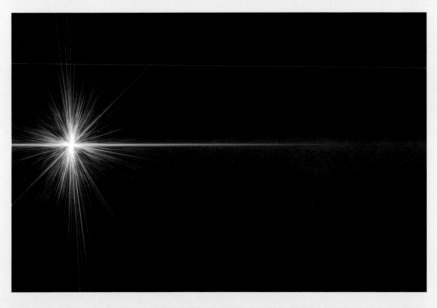

Before the mid-1930s, the only view of the Universe we had was the one we could see with our eyes. Even when enhanced with telescopes, we still only examined the objects that gave out visible light. That changed in the mid 1930s when an American engineer called Karl Jansky got involved in the development of radio telephones. It was Jansky's job to map out the sources of natural radio waves to ensure they did not interfere with communications. He discovered that radio waves were "shining" down from space. It appears that the stars gave out radio waves just like they gave out light.

The uptake of radio astronomy was slow due to the Second World War, but by the 1950s radio telescopes offered an entirely new view of the Universe. Much of the radio universe contained objects that were invisible to the eye, offering the first views of objects such as black holes, quasars and neutron stars. Many of the latter were seen as pulsars, or pulsating stars that emit a concentrated beam of radio as they rotate like a lighthouse. This appears to us as a flash of radio as the beam sweeps past us.

1960 The nuclear-powered submarine USS *Triton* completes a circumnavigation of the world completely submerged.

1961 Yuri Gagarin is the first person in space.
• The largest bomb ever exploded, *Tsar Bomba*, is detonated in the Siberian Arctic. It has a power of 50 megatons.

1962 NASA's *Mariner 2* probe flies past Venus becoming the first spacecraft to visit another planet.
• Neutrinos are shown to exist in several forms, or flavours.
• The SR-71 Blackbird spy plane enters service. It is the fastest jet-powered aircraft in history, capable of reaching 3,529.6 km/h (2,193.2 mph).

1963 Geophysicists prove that the seafloor is spreading out from the Mid-Atlantic Ridge.

1964 Murray Gell-Mann and George Zweig both propose that protons, neutrons, and other heavy subatomic particles are formed from groups of smaller particles called quarks.
• The cosmic microwave background is detected coming from the whole sky and gives the best proof so far of the Big Bang.

1965 Gordon Moore, a computer scientist, predicts that computing power will double every 18 months, a relationship now called Moore's law.

1966 Willi Hennig develops the idea of cladistics which classifies life according to evolutionary relationships.

QUASARS

In 1963, Maarten Schmidt, a Dutch astronomer working at Palomar Observatory, California, saw an object that was 2.5 billion light years away but appeared as a bright object in his telescope. That meant it was producing an unimaginably large amount of light. He calculated it was 4 trillion times brighter than the Sun – brighter even than the whole of the Milky Way. Schmidt called the object a quasi-stellar radio source, which has since become the word "quasar".

Other quasars were found and most were even further away, about 12 billion light years from Earth. That means we see them as light given out 12 billion years ago. The astronomy of the 1960s could not explain what a quasar was. One suggestion was it was the opposite of a black hole, a white hole that always gave out light. Another idea was that the quasar was actually huge star with a normal level of brightness, but its great mass bent space around it so much that astronomers were fooled into thinking it was billions of light years away. Both these ideas are impossible, however.

Instead, a quasar is an active galaxy where stars in the core are being swallowed up by a supermassive black hole. All galaxies are thought to be like this early on in their lives, which is why most quasars are seen far away – and back to the early days of the Universe. In 1998 a supermassive black hole was found at the centre of the Milky Way. It contains the mass of 4 million Suns and in the distant past must have blazed out light as it swallowed up stars.

Quasars give out a massive amount of light.

BACKGROUND SIGNAL

In 1964, Arno Penzias and Robert Wilson, two American astronomers, began using a supersensitive radio antenna. The Holmdel Horn Antenna in New Jersey was built by Bell Labs to test the first communication satellite technology. This used Echo "satelloons", which were large metallic balloons placed in orbit to act as space-based mirrors to reflect microwave communication signals. The two scientists found that the antenna detected a faint but persistent microwave signal and it was unchanging wherever they directed the device. They realized they had picked up a flash of energy that was predicted by the Big Bang theory. This energy had been redshifted, stretching its wavelengths longer for billions of years, and now the signal was not a bright hot light but a faint, cold microwave radio signal.

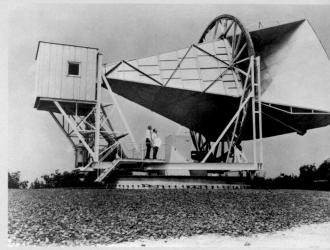

The Holmdel Horn Antenna.

Arno Penzias and Robert Wilson.

1967 The first pulsars, which are neutron stars spinning at huge speed and emitting beams of radio waves, are discovered by Jocelyn Bell Burnell in Cambridge.
• The first gamma-ray burst is observed; these are the brightest events in the Universe emitting more light in a few seconds than shines out from the Sun in its whole 10-billion year life
Lynn Margulis proposes endosymbiosis a theory that explains how cells of complex organisms, like plants and animals, arose from teams of bacteria working together.

1969 Neil Armstrong becomes the first person to walk on the Moon during the *Apollo 11* mission.
• String theory is proposed. This attempts to explain fundamental particles as vibrating strings and loops that move in extra spatial dimensions only valid at the quantum scale.
• The Internet is formed on connecting computer networks in California.

ENDOSYMBIOTIC THEORY

In 1967, the American geneticist Lynn Margulis set out an amazing theory as to why life is divided into two broad groups: the prokaryotes with their small unstructured cells, and the eukaryotes with their large cells with complex internal structures. The reason was quite amazing. The prokaryotes, made up of the bacteria and archaea, are the primordial life forms, with evidence of their existence stretching back at least 3.5 billion years. The first eukaryotic cells appeared around 1.5 billion years ago. Margulis suggested that happened when a set of different prokaryotes begin living as a team. Today these prokaryotic symbionts have become organelles. She showed that mitochondria evolved from purple bacteria – and the DNA in mitochondria back this up. The chloroplasts in plant cells, meanwhile, appear to be related to cyanobacteria, also called blue-green algae, which were the first photosynthetic organisms. This process of symbiogenesis would have been very rare, and all eukaryotes are descended from just one cell that achieved it successfully.

Lynn Margulis.

A VIEW OF THE PAST

The signal is now called the cosmic microwave background, or CMB for short. At the time of its discovery it offered the best evidence yet as to the veracity of the Big Bang theory, and all rival theories were shelved. The CMB is the release of energy that occurred around 380,000 years after the start of space and time. It took all that time for the Universe to cool down enough so that electrons could form neutral atoms by bonding to nuclei of protons. The amount of energy released was 100 times greater than all the energy coming from all the stars in the Universe. However, that vast energy is now spread across the whole of space and so is colder and less dense. The temperature of the CMB is just 2.7 K or –271°C

In the years since its discovery, the CMB has been mapped several times

REALLY SMOOTH

At first glance – and at many later ones too – the CMB appeared to be completely smooth or isotropic. In other words it looked the same no matter where you looked. However, maps of the CMB drawn up from the 1980s onwards, starting with the one made by the space-based observatory COBE (Cosmic Background Explorer) found tiny wrinkles. Where the energy density was a fraction higher, the

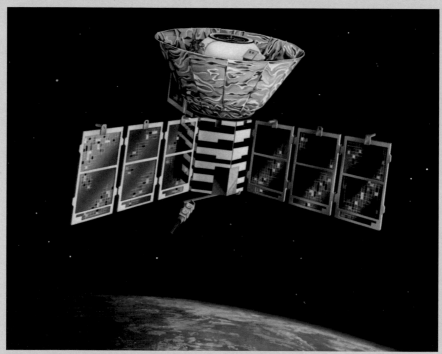

The Cosmic Background Explorer or COBE probe.

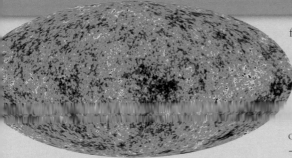

A map of the cosmic microwave background showing variations in temperature.

from Earth and space in ever greater detail to find clues to one of the biggest mysteries of all. Why has the Universe got anything in it? Why are there galaxies, stars and planets at all? The mathematics of the Big Bang shows us that for every particle of matter there is another of antimatter – and they both annihilate each other. However, the Universe is made of matter, with precious little antimatter to be seen.

galaxies formed in their billions. In places where the energy was reduced, huge voids of nothingness had formed.

This tallies with the 3D maps of the Universe being created today. These show that galaxies cluster together and galaxy clusters cluster into superclusters. Superclusters are the largest structures bound by gravity – our one contains 47,000 other galaxies – however, there are

bigger things out there. The universe has a filamentous structure, with superclusters lining up into sheets and "great walls" of stars that surround the galactic voids like the thin surface of a soap bubble. The Big Bang theory predicted the Universe should be empty, and most of it is. All of the stars we can see are packed into the thin filaments that formed from the wrinkles in the Cosmic Microwave Background.

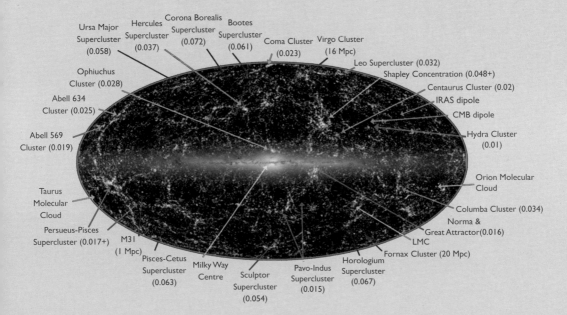

The Standard Model is the tool kit for the universe. It contains 17 fundamental particles that nature uses to create all mass, transfer all forces and contain all the energy in the universe. The ideas behind it emerged in the 1960s, and by the 1970s most of the particles theorized in the model had been observed "in the wild".

PARTICLE ACCELERATORS

The first addition to the Standard Model was the electron. This was discovered in 1897 using a vacuum tube and some magnets. Most of the other particles were discovered using particle accelerators, which also use vacuums and magnets, albeit on a much larger scale.

The idea for particle accelerators came from a trip in a hot air balloon in 1911. Researchers had found that the air became more conductive with altitude. To investigate, the Austrian Victor Hess set off in a balloon equipped with pre-charged electroscopes. He found that these devices has lost their charge as he increased in height. This showed that the air was more electrified and thus able to carry away a device's charge.

The source of this electrification came from space in the form of cosmic rays. These were streams of high-speed particles that flooded through space. They smashed into the outer atmosphere, knocking electrons off the atoms, and creating a layer of ionized gas at high altitude. Much of these rays come from the sun, of course, which means that the ionization is greater during the day and at night the electrified layer rises to a higher altitude.

The collisions of cosmic rays with air were the most high-energy particle collisions ever seen, and they revealed the first exotic short-lived particles within the Standard Model. Tiny blinks of Cherenkov radiation in the upper atmosphere provided evidence of fast-moving particles created in the collisions.

Victor Hess prepares for his 1911 air balloon flight.

1970 The Boeing 747, or Jumbo Jet, enters service as the first double-decker, wide-body passenger jet.
• Liquid crystal displays or LCDs are patented and used in early digital watches and similar displays.

1971 The first space station, *Salyut 1*, is put in orbit around Earth. However, the crew could not enter and all three died on the way back to Earth when their capsule depressurized. These are the only deaths to occur outside the Earth's atmosphere. The Soviet space station programme continues through the 1970s

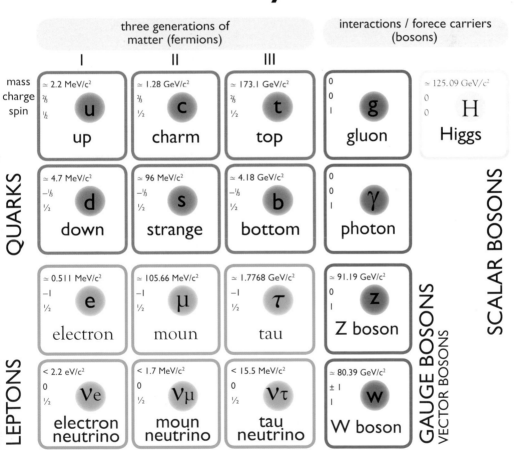

Standard Model of Elementary Particles

The Standard Model describes the Universe in 17 fundamental particles.

EXOTIC PARTICLES

In 1934, Hideki Yukawa suggested that there were particles that were halfway between electrons and protons. He named them mesons. In 1936, the mu meson was discovered in cosmic-ray collisions. However, this did not match Yukawa's theory and was later renamed as a muon. It is a kind of oversized and very short-lived relative of the electron. In 1947, a pi meson (or pion) was seen in jets of particles spraying out from cosmic ray collisions. It had two-thirds of the mass of a proton and was the first true meson. Although it only exists for a few billionths of a second, the pion and other mesons like it, with their strange fractional mass, provided the first clues that they and the protons and neutrons were not actually fundamental particles. Perhaps they were made from even smaller units?

To find out, physicists were going to need collisions with even greater energy than seen in cosmic rays and be able to witness collisions up close. For that they needed a particle accelerator. The pioneer in this field was the American Ernest Lawrence. In 1929, he had invented the cyclotron, a device that sent ions and other charged particles around on a spiral course set by oscillating electric fields. The particles sped up as they reached the centre and smashed into a target. A similar use of magnetic fields to focus particles and electric ones to drive them along is used in larger linear and looped accelerators, growing even larger and more powerful.

The collisions they created sent out sprays of particles, and the paths of those particles were tracked in cloud and bubble chambers. Streaks of bubbles or condensation were photographed and magnetic and electric fields in the detectors helped to indicate the mass and charge of each particle.

Hideki Yukawa.

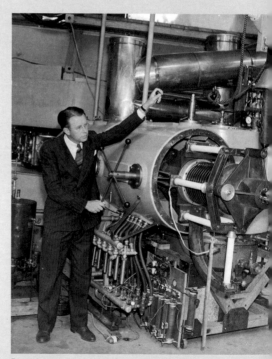

Ernest Lawrence with a cyclotron, an early form of particle accelerator.

1972 The first genetically engineered strand of DNA is produced.

• James Lovelock presents his Gaia hypothesis to the public. His theory describes how Earth is a self-limiting entity that uses feedback to control the conditions in the biosphere.

1973 NASA launches *Skylab*, an orbital laboratory built from a modified Saturn V rocket. It is the first and last US-only space station.

1974 Sagittarius A*, a powerful radio source, is found at the centre of the Milky Way galaxy. It is later found to be a supermassive black hole weighing 4 million times as much as the Sun.

• The British physicist Stephen Hawking combines the theories of quantum physics and general relativity to investigate black holes. He proposes that even black holes radiate energy and thus gradually lose their mass.

• Rudolf Jaenisch creates the first genetically engineered organisms.

1975 The Soviet *Venera 9* is the first spacecraft to make a successful landing on another planet. It sends back the first pictures from the surface of Venus.

1976 *Viking 1* is the first probe to land on Mars. It sends back colour pictures of the red planet and analyzes soil samples for signs of life.

1977 Voyager 1 and 2 are launched on a grand tour of the outer planets. Both visit Jupiter and Mars, while *Voyager 2* flies on to Uranus and Neptune.

• The magnetic resonance imaging machine, better known as MRI, uses radio emissions from powerful magnets to make images of internal soft body parts.

1978 Charon, the largest satellite of Pluto, is discovered.

1979 The first magnetar, a magnetic neutron star, is discovered.

• *Seawise Giant*, the largest ship ever made, is launched as an oil tanker.

1980 Cosmic inflation is proposed by Alan Guth to explain the expansion of the early Universe.

CENTRAL DOGMA OF MOLECULAR BIOLOGY

mRNA from the nucleus is "read" by a ribosome.

Following the triumphant discovery of the double helix structure of DNA in 1953, biochemists wanted to know how the chemical could carry a genetic code. That investigation took the best part of the 20 years, and the process it revealed is now called the Central Dogma of Molecular Biology. If DNA is shaped like a twisted ladder then the rungs are built from pairs of molecules called bases. They are four: Adenine (A), thymine (T), cytosine (C), and guanine (G), and the sequence of these bases in larger DNA molecules spells out a code of four letters, A, T, C and G. This code carries the information for making the many thousands of different kinds of proteins a living cell and body needs. Proteins are strings of smaller units called amino acids. A unit of three base "letters" codes for one amino acid, and a gene contains code for all the amino acids in one protein strand – as many as a 100 amino acids. The genetic code is stored in the cell's nucleus locked within a double strand of DNA. To turn it into a protein, the gene is copied on to a single strand of mRNA (messenger RNA). This leaves the nucleus and is fed through a ribosome, which connects amino acids one by one in the right order until they make a functional protein. The dogma of the process is that information can only travel from the DNA to the protein, never in the other direction.

DARK MATTER

One thing – and it's a big thing – that the Standard Model does not cover is dark matter. Dark matter is material that interacts with the force of gravity but does not interact with electromagnetism. So, it makes no light, it reflects nothing, and simply cannot be seen. All that is detected is its gravitational effect on the ordinary matter around it, and this is where the first hint of it was revealed. In the 1930s, astronomers were avidly observing galaxies, only freshly revealed as island universes in their own right far beyond the shores of our galaxy. Astronomers measured how heavy a galaxy was – how much matter it contained – from the intensity of its light. Oddly, galaxies seemed to rotate faster than their measured mass suggested. If the galaxy really was as lightweight as observed, its internal gravity would not be strong enough to hold it together. As it spun at great speed, its stars would be flung out to all corners of the Universe. Astronomers mostly assumed they'd made an error in the measurements and went on to answer the easier questions about galaxies. Fritz Zwicky noted that there might be some dunkle Materie (dark matter; he spoke German) inside and that was making the galaxies heavier.

It was not until 1979 that Vera Rubin, an astronomer working in Washington, DC, began to measure the rotation of the Andromeda galaxy. She did so by taking the redshifts and blueshifts of stars swinging away and towards us to produce a very precise result. She confirmed the mismatch in speed and mass, and was able to calculate that the ratio of matter to dark matter is about 1:6.

Vera Rubin

What dark matter is or represents remains a mystery. Perhaps it is a set of hidden particles that are beyond the current version of the Standard Model. It could be, at least in part, that ordinary matter is spread through interstellar space in larger quantities than previously thought. A third option is that the gravitational effects seen as dark matter are the result of ordinary matter in alternative universes that warp the space–time of this one. Or it shows that the Standard Model and relativity and just about everything else in physics is wrong.

The Andromeda galaxy.

QUARKS

In 1964, two American physicists, Murray Gell-Mann and George Zweig, working independently proposed that proton and neutrons and the mesons were made from smaller particles. Gell-Mann named them "quarks", a word taken from a book by the Irish writer James Joyce. Neutrons and protons have three quarks, while the mesons had just two. Any body made of quarks is now known as a hadron. In 2015, the Large Hadron Collider particle accelerator in Switzerland reported making a pentaquark, a hadron with five quarks in it.

There are six types, or flavours, of quark: up, down, top, bottom, strange, and charm. The most common are the up and down, and the others eventually break down into them. Each quark has an electrical charge of either a third or two-thirds. A proton has two up and one down quark and a charge of +1, while a neutron has one up and two downs and an overall

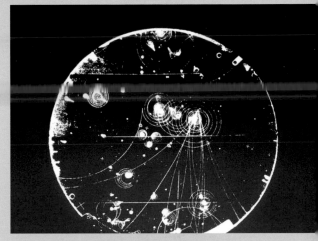
The paths of particles are tracked in a bubble chamber.

charge of 0. Combinations of three quarks always result in a whole number charge ranging and there are exotic and short-lived combinations with charges ranging from −1 to +2.

The first observational evidence of quarks came in 1969, and the last one, the top quark, was found in 1995. This amazing particle is the size of an electron – effectively dimensionless – but weighs more than the first two-thirds of the atoms in the periodic table!

Quarks are held in place by the strong nuclear force, which also makes the atomic nucleus stay together. The Standard Model includes bosons or the particles that mediate fundamental forces. The gluon controls the strong force, and the photon controls the electromagnetic force. The weak nuclear force, which is involved in pushing items out of a nucleus during radioactive decay, is controlled by the W and Z bosons. Only gravity lacks a boson in the Standard Model. To find it, particle physicists would first need to understand how gravity works as a quantum level, and at the moment that is a mystery.

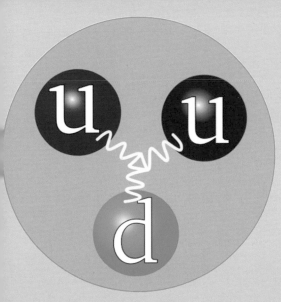
The three quarks, two up and one down, inside a proton.

NEUTRINOS

Another piece of the Standard Model are the neutrinos, the smallest particles of all. As the name suggests, they are neutral, and there is one for each member of the electron family, which also include the muon and tau (which is even more unstable than the muon.) They are produced by radioactive decay where protons and neutrons switch from one to the other. This kind of thing is happening all the time during nuclear fusion, so the Universe is teeming with neutrinos that stream from stars including our Sun.

There are 65 billion neutrinos passing through every square centimetre of the Earth's surface every second. They are unaffected by electromagnetic fields and so just pass through. They do respond to gravity but being so lightweight – half a million of them add up to the weight of an electron – they are very hard to find.

The seventeenth and last member to be added to the Standard Model was also very hard to find, and required the largest machine ever devised. Find out more about that on page 234.

A neutrino detector.

Alan Guth.

COSMIC INFLATION

The Big Bang Theory is pretty watertight but by 1980 there were a few big holes that needed filling. The first was that the theory dictated that the fundamental forces separated out into distinct interactions after the barest fraction of the first second of the Universe's existence. At that time the Universe would have been very small and very hot and that energy density would create strange effects. These effects did not appear, so the Universe must have been bigger and colder by the time the forces decoupled. Secondly, the expansion of the Universe is very flat, meaning it is extending equally in all directions. For that to occur its energy must be distributed very smoothly. How did that happen? Lastly, light from the edge of the Universe has taken the entire lifetime of the Universe to reach us. So when the Universe was 10 minutes old, the furthest you would have been able to see would be 10 light minutes away (a bit further than the distance to the Sun). After a year, the edge of the Universe was one light year away, and so on. Because the Universe is always getting bigger, there has never been enough time for light to travel from one edge of the universe to the other edge. No information or energy has ever been exchanged between these opposite ends of the Universe. However, when you look at them they both appear to be structured in the same way. How come?

In 1981, the American Alan Guth developed the theory of cosmic inflation and added it as an early stage to the development of the Universe to plug these holes in the Big Bang Theory. According to Guth, the young Universe expanded many times faster than the speed of light during the first 10^{-35} seconds of existence (that's 100 billionths of a septillionth of a second.) In that time the Universe doubled in volume at least 100 times, swelling up from a dimensionless point in space to the size of a marble.

This rapid expansion cooled the Universe faster, thus solving the forces problem. The young Universe grew so fast that everywhere that is close enough to be observable from one point (such as the Earth) at some point in its history all began in the same local bit of the Universe, so little wonder it all looks the same. Finally, inflation ended when energy became smoothed out enough. Tiny fluctuations remained and these were locked in as the structure of the Universe for the rest of time. (The theory predicts that inflation keeps on going for ever, bubbling off new universes as it goes. What hope there is for proving the existence of multiple universes outside our own is a question with no answer in sight.) Inflation fits well with the rest of the Big Bang Theory but there is still no direct evidence for it.

Electrical resistance is the biggest cause of energy loss in our everyday appliances, like computers, TVs and the electricity grid. A little over 100 years ago the phenomenon of superconductivity was discovered, in which a material has an electrical resistance of zero. As well as being a way to examine tiny quantum effects on the large scale, this discovery raised the prospect of a future technology of near

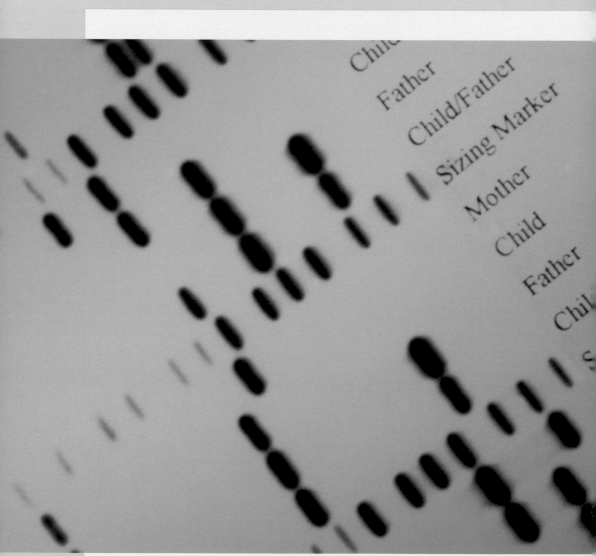

DNA "fingerprints".

total efficiency. However, there was a snag. Superconductivity was only possible at very low temperatures. The energy needed to reach these low temperatures negated any efficiency savings.

DNA FINGERPRINTING

No TV crime drama is complete today without a mention of DNA samples that point to a suspect. And true enough, DNA evidence has led to convictions today that were out of the reach of investigators before the mid 1980s – it is also being used to secure convictions of historic crimes dating back 50 years or more. This is all thanks to the DNA profile, or "fingerprinting" technique developed by the British geneticist Alec Jeffreys at Leicester University in 1984.

The DNA profile is not a complete record of a person's genes. That would have limited the use as a form of identity because around 99.9 per cent of human DNA sequences are the same in every person. Instead, Jeffreys's system uses the cell's own machinery to look for and multiply sections of DNA code that are repeated at least twice, and perhaps many more times. These sections are amplified, dyed and then separated on a gel to reveal a pattern of bands of different thicknesses, each one indicating a region of repeated code. Statistically, there are many areas where code repeats in every person, but the number of repeats of different types is highly individual, although not unique. The pattern of repeats runs in families, and so the fingerprint can be used to prove familial relationships with great certainty.

1980 Luis and Walter Alvarez present evidence from rocks 69 million years old that the dinosaurs were made extinct by the impact of a giant meteorite.

1981 NASA's space shuttle *Columbia* makes its maiden flight and becomes the first reusable spacecraft.

1982 On March 10, all nine planets (including Pluto) align on the same side of the Sun in an event called syzygy.

1983 The F-117 Nighthawk is the first stealth plane to enter service. It has some ability to avoid detection by radar but is not very manoeuvrable and is very hard to fly.

1984 HIV, the Human Immunodeficiency Virus, is identified as the cause of the Acquired Immune Deficiency Syndrome or AIDS.
• Alec Jeffreys develops DNA fingerprinting at Leicester University.

1985 A new form of carbon is discovered made from 60 atoms arranged in a ball. The substance is called buckminsterfullerene.

1986 The *Challenger* space shuttle is destroyed on launch, killing all seven crew.
• The ESA's *Giotto* and several other probes fly past Halley's Comet on its most recent visit to the Earth.
• Superstring theory is proposed in an attempt to unify all particles.

1987 SN 1987A becomes the first supernova witnessed by modern astronomers.

1989 Stanley Pons and Martin Fleischmann announce cold fusion at the University of Utah, promising a limitless supply of cheap energy. However, the pair's results are proven to be overblown and incorrect.

SUPERCONDUCTORS

TEMPERATURE DEPENDENT

Nevertheless, significant technologies have been developed the use of superconductor. These include MRI, magnetic resonance imaging, which can take detailed pictures of soft tissues in the body. Mass spectrometers use powerful, superconducting magnets to bend beams of charged particles and molecules as a means of measuring their size and chemical makeup. Finally maglev, or magnetic levitation, technology uses hugely powerful superconducting magnets to have an entire train float above a magnetized track and move along at record-breaking speeds. All these technologies need to have components working below −260°C.

In 1986, the engineers Karl Müller and Johannes Bednorz developed a ceramic that could superconduct at only −163°C. This may not seem very warm, but in the parlance of superconductors anything warmer than the boiling point of liquid nitrogen (−196°C) is seen as "high" and therefore relatively easy to maintain. (Liquid nitrogen is a relatively cheap and plentiful source of cold.)

Research into high-temperature superconductors continues, with some materials are reported to work at just −70°C (although under very high pressure). The search continues for the holy grail of the room-temperature superconductors. Despite the name, these are classified to work at 0°C (that's a cold room), but engineers would be more than happy to be able to cool these materials with everyday freezer technology.

A levitating magnet requires superconduction.

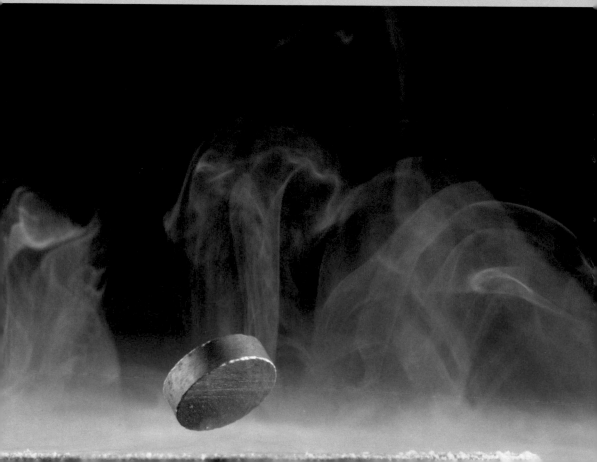

The rapid transit to Shanghai's airport is a maglev train.

DEEP COLD

It was the rise of refrigeration technology that led to the discovery of superconductors in the first place. In 1898, the reportedly objectionable Sir James Dewar had succeeded in cooling hydrogen gas to –253°C, making it condense into a liquid. The following year he was able to freeze it solid. He had assumed that hydrogen, the smallest and simplest of elements, would have the lowest melting-point in nature. To his dismay, research into helium showed that it did not condense until –269°C! Due to his reported objectionable nature, Dewar was unable to secure enough helium (very rare at the time) to succeed in liquefying it, and lost his place in that particular chapter of the history books (he's in others).

That honour fell to the Dutch researcher Heike Kamerlingh Onnes, who succeeded in liquefying helium in 1908. He then began using the liquid helium as a coolant and investigated what happened to other materials when you chilled them to a few degrees above absolute zero. In 1911, Onnes cooled mercury to around 4 K and he found the metal's electrical resistance disappeared entirely. It had become a superconductor. In 1933 it

Heike Kamerlingh Onnes.

was also found that when it becomes a superconductor, it also blocks every magnetic field from entering. This effect is behind the phenomenon of magnetic levitation.

Electrical resistance is caused by atoms getting in the way of the flow of electrons that form an electric current. The atoms form more of a barrier as they are jiggling around at high temperatures, so most materials increase in resistance as they get hotter. Removing heat energy slows the material so it offers less resistance. However, why resistance disappears completely is still a mystery. The best theory is Cooper pairs, which are two electrons that couple together at low temperatures. Working together, the electrons are not scattered around as they move through a material, like a single electron is, and so they do not lose energy through resistance.

219

Alongside the communications revolution, a quieter shift in technology was taking place in the life sciences in the 1990s. In 1996, Dolly the sheep was the first mammal to be cloned as the genetically identical daughter of her mother, and stem cell technology began looking at how body cells could be programmed to grow new body parts. Genetic modification techniques were being developed that could be used to introduce new genes into the DNA of any living thing. The most notable system was CRISPR (an acronym for something even more arcane – Clustered Regularly Interspaced Short Palindromic Repeats). CRISPR is a set of

Dolly the sheep.

DNA codes used by bacteria to cut away unwanted DNA introduced by viruses. Genetic engineers have taken control of this ability and use it to cut out DNA from any organism – and that gene can

The genetic inheritance – or pedigree – of domestic animals is very important.

then be added to the DNA of an entirely different species. Today CRISPR and related technologies are an everyday tool capable of manipulating DNA on a large scale. Genetics is still catching up with the technology as scientists figure out what gene does what – and whether they are worth using in genetic modification. However, it is certain that this biotechnology will play a large part in the future of medicine and agriculture. Whether society allows it to be used for cosmetic, aesthetic and social purposes remains to be seen.

1990 The Magellan orbiter uses radar to make a detailed map of the surface of Venus, which is hidden from view by thick clouds.

• The Hubble Space Telescope is launched but needs to be optically corrected because of a misshapen mirror.

• Carl Woese reorganizes the classification of life into three domains, bacteria, archaea and eukaryotes, the latter of which includes all complex life.

REORGANIZING LIFE

Carl Linnaeus, in creating his system for classifying life, divided organisms into two great kingdoms: Animalia and Plantae. Today most biologists work with five kingdoms, by adding Monera (the bacteria), Protista (the amoebas and such), and Fungi (the mushrooms and moulds). However, that was not enough for the American biologist Carl Woese. In 1977 he found that the Monera was actually two distinct genetic groups. One of them corresponded with what we understand as bacteria, but the other contained organisms that were extremophiles, meaning they survived in conditions that would kill all other cells. Woese named these life forms Archaea. They are mostly found in hot volcanic springs or in deep rocks bathed in salts and other chemicals. In 1990 Woese proposed that life should be reorganized into three domains which out-ranked the kingdoms: Archaea, Bacteria, and Eukarya. Woese organized life this way after showing that members of each domain are set apart by having a distinguishing form of ribosomal RNA (rRNA) within their cells. In 2012 the Swedish microbiologist Stefan Luketa proposed adding two more domains, which included viruses (infectious DNA and protein structures) and prions (infectious proteins, lacking DNA).

Carl Woese.

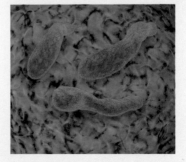
Archaea.

ARTIFICIAL SELECTION

Humans have been modifying the genes of plants and animals for millennia though artificial selection. Breeders selected which domestic animals or crop plants were used to breed the next generation. They always chose the ones that grew fastest and produced the best food. In so doing the farmers were altering the gene pool, or mix of genes that a population of organisms had.

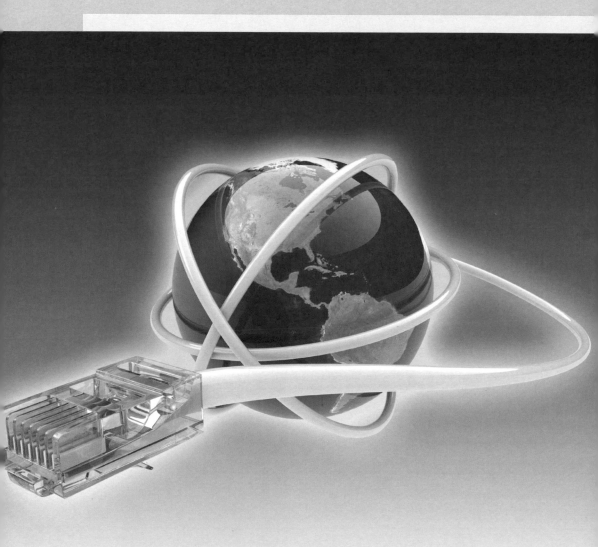

1991 Buckminsterfullerenes are reassembled into long nanotubes, making superstrong, superlight, superconducting materials.
• The lithium-ion battery becomes commercially available after a decade of development

• Nexus, the first web browser, is developed by Tim Berners-Lee
1992 The Cosmic Background Explorer finds tiny anomalies in the temperature of the Cosmic Microwave Background.

The World Wide Web

The 1990s was the decade of the World Wide Web, an exciting time before the dominance of Google and Facebook. In 1940, George Stibitz connected an electromechanical calculator to a telephone line and was able to operate it using a keyboard at the other end of the line. This was the first networked computer, something large organizations were to come to rely on. By the 1960s the US military were developing a system to protect their communication network from attack. The result was ARPANET, the Advanced Research Projects Agency Network. This network used a system called packet switching, where every signal was broken up into little packs which made their own way from the sender to the receiver and were reassembled at the other end. Packets found their own route through the network, and if one of them did not arrive, the receiver sent a message back to request it again. This system made networks very robust against attacks. Even if some lines of communication were cut, the packet switching would automatically get the message through.

ARPANET was switched on in 1969 and began to connect the networks of several universities, first across the United States and then the world. Then the business world and communications industry plugged in and the fast growing network of networks became known as the Internet.

To many, the terms "internet" and "web" are used to mean much the same thing. However, the Internet is the physical network of computers and cabling, while the web is a system for sharing information. It was invented in 1989 by the Briton Tim Berners-Lee, a computer scientist working at the CERN research centre near Geneva, Switzerland. The idea was to view the contents of another computer (the bits made public anyway) through a window on your computer. That way only the bits you needed right there and then were sent through the Internet, not the whole set of files. Berners-Lee developed HTML, or Hypertext Markup Language, which arranged words and pictures and allowed users to click on links between one document and the next. It is all so ubiquitous now it is surprising to the think that the "web browser" software most of us use every day to look at the news, buy stuff and catch up with friends, was only put on general release by Berners-Lee in 1991.

GENETIC ENGINEER

Selective breeding is very slow, especially for long-lived organisms like fruit trees. By the time scientists understood how genes worked, they began to manipulate them by taking control of the cellular mechanisms that copied, read and kept DNA organized in the nucleus. In 1972, Paul Berg managed to combine the DNA of two viruses to make an entirely new, or transgenic, entity. Two years later, Rudolf Jaenisch created transgenic mice by adding DNA of a retrovirus into their cells while they were still embryos.

Genetic engineers began using similar techniques to enhance crops, adding genes from other organisms to help them be more resistant to frost and disease. The process of making genetically modified plants involves editing out the gene to be added from its original donor organism. Then this gene is either inserted as part of a virus, which enters a target cell and introduces the DNA, or a gene gun can be used to blast cells with vast numbers of the DNA bound around tiny particles of gold. Most of the DNA and cells are simply splattered, but in a few cases, the DNA enters a cell intact and is incorporated into the material of the cell's nucleus.

In some countries people are suspicious of genetically modified organisms (GMOs), especially crops and farm animals. They worry that these transgenic organisms will escape and their artificial genes will be incorporated into wild gene pools, the consequences of which are untested and unknown. However, there are many other ways in which GMOs are being used. For example, the insulin used to treat diabetics is mass-produced in a very pure and safe form, using GM bacteria that carry the human insulin gene. Although one bacterium produces only a tiny

Rudolf Jaenisch.

volume of insulin, billions of bacteria can produce a limitless supply.

A clone is an individual that is an exact genetic copy of another original organism. Many plants and small animals,

Gene gun.

like aphids, effectively reproduce by cloning themselves. Its very fast but also risky. Sexual reproduction mixes genes to make a varied population, which is more robust against changes in conditions, especially disease. However, genetic engineers wanted to investigate cloning as a way of creating GMOs by bypassing the usual breeding process.

1994 The *Galileo* probe en route to Jupiter takes pictures of Comet Shoemaker Levy 9 colliding with the giant planet.
• The first stage of the Global Positioning System is completed and plans start to make it publicly available.
• The Solar and Heliospheric Observatory, or SOHO, is launched and takes up an orbit of fixed point between Earth and the Sun. This allows it to image the Sun without Earth getting in the way.

GLOBAL POSITIONING

Today we would be lost without our car satnav or smart phone map app – quite literally. Where once we consulted printed maps, we now rely on a computer to tell us where to go, and for that the device needs to know where we are right now. That is achieved using GPS or the Global Positioning System. Initially developed as a navigation tool by the US military, GPS uses a constellation of satellites to triangulate positions. After 20 years of work, the 24th GPS satellite reached orbit in 1994 making the system fully functional. By 1996 President Bill Clinton made the full version of GPS available to the general public, so everyone could pinpoint their location to within a few metres. Today the system works with 31 satellites, and similar systems are in development by Russia, China, and the European Union.

For GPS to work, a device like a satnav needs to pick up signals from at least three satellites. The signals broadcast the position of the satellite and the time of the signal. The satnav compares the time difference between the sending and receiving of the signal to calculate a distance to each satellite. It then uses those three distances (often more) to figure out its exact position, anywhere on the surface of the globe.

The ubiquitous car satnav.

DOLLY THE SHEEP

In 1996, a sheep called Dolly was the first cloned mammal to be born. Clones are often said to have one parent, but really Dolly had three. A cell was taken from the udder of Dolly's biological mother. The nucleus was removed from the ovum, or egg, of another female sheep and replaced with the nucleus from Dolly's first mother. The modified egg was then given a tiny electric shock to make it start dividing into a ball of cells. This embryo was placed in a third sheep, Dolly's surrogate mother.

Dolly was then born in the normal way at the Roslin Institute, Edinburgh, Scotland.

However, she was the only survivor of 227 techniques, and so cloning technology has not proved to be a game-changer. Therapeutic cloning is proving to be more useful. This aims to make copies of embryonic stem cells, not an entire individual. Stem cells are the foundation stones of a body and if the technology can be cracked then it could be used to grow new body parts or repair injuries.

THE HUMAN GENOME PROJECT

A genome is the complete sequence of DNA stored in an animal's cells. In 1990, a project began to decode every last letter of the human genome. The first draft of the Human Genome Project was published in 2001. It is made up of 3 billion letters, one of A, C, T or G. That is enough to fill 200 telephone directories when printed as readable type. Every cell nucleus holds a complete set of the genome. If all the DNA used in just one cell to encode it was unravelled, it would be 2 m (6.5 ft) long! While every human has a unique genetic code, we all have the same set of genes (just a unique set of alleles or version of genes.) The DNA used by the Human Genome Project came mostly from one anonymous man from Buffalo, New York.

The project tells us that humans have around 20,000 protein-coding genes, which is way below the original prediction of 100,000. Today geneticists are using the genome data to identify genes. We are still in the dark about what most of them do for us. It turns out the decade-long job of decoding the genome was the easy bit.

1995 The top quark is discovered, the last and heaviest of the six flavours.

• The first Bose–Einstein condensate is produced, a material so cold that it forms a fifth state of matter where quantum effects appear on the large scale.

• The Digital Versatile Disk, or DVD, is standardized as the high-capacity optical disk with enough space to hold a feature film.

1996 NASA scientists report finding primitive bacteria-like forms in a meteorite found in Antarctica and originally from Mars. It is suggested they are fossil life forms but that assertion has since lost favour.

• Dolly the sheep is the first mammal to be cloned.

1998 Researchers attempting to measure how quickly the expansion of the Universe is slowing are shocked to discover that the rate of expansion is actually speeding up. This acceleration is put down to a mysterious anti-gravity dubbed "dark energy."

• Zarya, the first module of the International Space Station, is launched. The ISS is now the largest spacecraft in history.

1999 The term Wi-Fi is introduced for wireless radio connections to the Internet.

2000 Life scientists Peter Ward and Donald Brownlee propose the Rare Earth Hypothesis, arguing that Earth's complex life-forms and the evolved intelligence of humans is only possible due to several characteristics of Earth that would be highly unlikely to be repeated elsewhere in the Universe.

• The crew of *Expedition 1* arrives on the International Space Station. The ISS has had a human crew on board continuously ever since.

• The human genetic code is cracked.

The cloning process used to create Dolly the sheep in 1996.

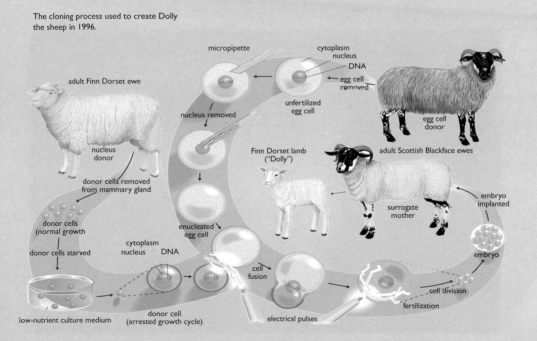

adult Finn Dorset ewe

micropipette

cytoplasm
nucleus
DNA
egg cell removed

unfertilized egg cell

nucleus removed

nucleus donor

donor cells removed from mammary gland

donor cells (normal growth)

donor cells starved

cytoplasm nucleus DNA

enucleated egg cell

Finn Dorset lamb ("Dolly")

egg cell donor

adult Scottish Blackface ewes

surrogate mother

embryo implanted

embryo

cell division

fertilization

low-nutrient culture medium

donor cell (arrested growth cycle)

cell fusion

electrical pulses

The Big Bang theory arose primarily from the observation that the Universe is expanding. If time were made to run backwards, the Universe would be seen to shrink, and so at some finite point in the distant past, the entire Universe occupied a minute, dimensionless volume. Start the clock running forward again, and the Universe erupts in a Big Bang, a tumultuous event that happened throughout the entire Universe all at once — only the Universe was very small indeed back then.

By the turn of the 20th century there was a huge amount of evidence supporting the Big Bang theory. The most significant was the cosmic microwave background (described on pages 206-7), which is a faint glow throughout the Universe, left by the almighty flash of energy released when matter formed into atoms for the first time. The distribution of energy in the CMB was also being found to match the distribution of matter — the galaxies and stars — observed across the Universe.

BIG CRUNCH

A big question remained, however. Would the Universe keep expanding forever, or would gravity apply the brakes at some point — and if so, would the contents of the Universe hang in at this maximal volume or would it pull itself together again?

The first scenario was described as an open universe. If the Universe was indeed open it was predicted to result in Heat Death in the far future — hundreds of billions of years hence. This would entail all the matter in the Universe becoming so spread out and so cooled down that

The Big Bang (perhaps).

GRID CELLS

In 2005 Edvard Moser and May-Britt Moser discovered how the brain builds a model of the space around it using layer of grid cells. Each grid cell is like a pixel on a unit, and together they create a representation of space like a shape plotted on graph paper. The cells actually form a triangular network, not a square one, but the effect is believed to be the same. The grind cells are the site of spatial memory. If all outside stimuli are removed, the brain consults the grid cells to figure out where it is.

all physical interactions would end. The latter situation – a closed Universe – would have an outcome dubbed "the Big Crunch". This would be almighty collision of all the matter in the Universe, perhaps leaving nothing but an immense black hole surrounded by an infinitely dark and empty space.

Whether the Universe was open or closed, one of the central tenets of the Big Bang theory was that expansion of space–time was slowing down. Being able to measure that deceleration would not only reveal whether the universe was open or closed – and thus settle the ultimate fate of the Universe – it would also offer a chance to "weigh" the Universe. Cosmologists were able to predict the amount of regular matter that was in the Universe, but were still less sure about how much dark matter was out there. Dark matter was a mysterious substance that produced gravity but was completely unaffected by light or magnetism – and therefore was proving near impossible to observe. If it was possible to measure the deceleration of space–time, then it would give a clear figure of how much matter – or more correctly, how much energy – was in the Universe.

2001 Near Earth Asteroid Rendezvous (NEAR) Shoemaker lands on Eros, the first touchdown on an asteroid.
• Neutrinos are discovered to have mass.
2002 Quaoar, a dwarf planet in the Kuiper Belt, is discovered.
• Grigori Perelman solves the Poincaré conjecture, a century-old mathematical problem.
2003 China is the third country in the world to send an astronaut into space.
2004 The Huygens lander finds lakes of gasoline on Titan, the largest moon of Saturn.
• The quantum state of one atom is transferred – or teleported – to another for the first time, making use of the phenomenon of quantum entanglement.
• Social networking internet site Facebook is founded.
• *SpaceShipOne*, a rocket-powered spaceplane, is the first craft to fly into space twice in a fortnight, thus securing the $10 million Ansari X Prize.
2005 Edvard and May-Britt Moser identify the cells that the brain uses to work out spatial positioning.
2006 Pluto and Ceres are reclassified as dwarf planets along with several large bodies found in the Kuiper Belt and Oort Cloud.
• New Horizons is launched with a rendezvous with Pluto in 2015.
• The first synthetic bacteria are produced using DNA assembled in a laboratory.
• The Palm Jumeirah, an artificial island in the shape of a palm frond, is completed adding 40 km (25 miles) to the coastline of Dubai.
2007 Apple iPhone is introduced.
• Amazon Kindle is released.
2008 The field of epigenetics becomes significant when it is discovered that chemical changes in the structures of chromosomes can be inherited by at least two generations.
• Construction of the Large Hadron Collider, the largest particle accelerator in history, is completed.

SPEED TESTING SPACE

In the mid 1990s a method for speed testing space became apparent. A team of astronomers working in Chile had been able to prove that a type 1a supernova was a standard candle. This arcane language meant that the distance from Earth to a particular type of supernova explosion that was seen in relatively large numbers beyond our galaxy could be calculated from the brightness of the explosion as seen from Earth. A type 1a supernova is produced in a very particular way. A white dwarf star is in a binary system with a larger, giant star. The white dwarf pulls material from its giant neighbour, steadily increasing in size — growing from the diameter of Earth to an object 1.38 times heavier than our Sun. At this precise size, the star explodes. All these explosions are produced by stars of the same size and brightness, so if one appears dim it is further away than another that appears brighter.

Each explosion also had a redshift. This is one of the most significant tools in astronomy. As the light from any astronomical object pours through expanding space, the light wave is stretched, increasing its wavelength and lowering the frequency. This is termed as a redshift, because the expected colours

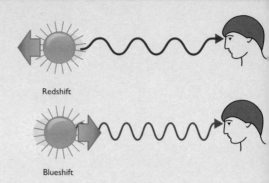

Redshift

Blueshift

of light are all shifted towards the lower frequency red end of the spectrum.

Redshift is a means of telling astronomers how fast an object is moving away from the Earth. (Beyond our galaxy, everything is moving away.) However, another tool of astronomy is light time: it takes longer for the light from a distant object to reach Earth than the signal from a nearer object. For example, it takes eight minutes for light to arrive from the Sun, and four years for it to get here from our nearest stellar neighbour, the Alpha Centauri system. As a result, we see what the Sun looked like eight minutes ago. If it went out, we'd be none the wiser for eight minutes. Meanwhile, the stars of Alpha

The first X-ray images of a type Ia supernova provided observational evidence that this is the explosion of a white dwarf orbiting a red giant star.

Centauri could have disappeared last year (although they are not likely to have) and they would only blink out of the sky three years hence.

2009 Successful gene therapies for medical problems are introduced.
• The Kepler space observatory is launched by NASA.
2010 Element 117 is produced in Russia, and makes the final member of the seventh period, or row, in the periodic table. It is later named Tennessine.

KEPLER TELESCOPE

The first extrasolar planets were discovered in the 1990s, showing that Earth and its family of planets were not a special case in the Universe. However, spotting planets orbiting distant stars from Earth's surface was a very difficult process. In 2009, the Kepler space observatory was launched to do it from outside the atmosphere. Focusing on a relatively small patch of sky, the telescope looked for small fluctuations in the brightness of stars that indicated that planets were orbiting in front of them, blocking out their light. Kepler has identified more than 4,000 such stars, and sensitive instruments on the surface are now working through each candidate looking for tiny redshifts that show how the stars are made to wobble by the gravity of their planets. So far more than 2,300 exoplanets have been confirmed, and this suggests that planets are probably more numerous than stars. The next question is: are any of these alien planets like Earth?

The Kepler spacecraft was finally decomissioned in 2018.

MEASURING THE UNIVERSE

In the late 1990s, two teams of researchers, one based in the United States, the other in Australia, decided to combine type 1a supernova, redshift and light time to measure the rate of expansion of the Universe. The teams searched out type 1a supernovas — often taking it in turns to use the same telescope in Chile — and then measured their redshifts using another telescope in Hawaii. They followed a hectic schedule to make the most of their telescope time. The brightness of the stars told researchers how far away they were. The redshifts indicated how fast the stars were moving (due to universal expansion) and the light-time revealed how old the stars were — many were shining out from billions of years in the past.

The general idea of both research teams was that distant objects, which were being observed from a time when the Universe was young, would be moving very fast. They represented a time when the young Universe was expanding faster than it is now — according to the theory. Nearer objects, which represented the older part of the Universe, were predicted to be moving slower due to the expansion of space and time slowing down.

However, both teams found something so incredible they felt sure it was a mistake — until they confirmed to each other that their results matched. About 7 billion years ago, the expansion of the Universe started to speed up, not slow down. When they crunched the numbers using the old theories, the data showed that the Universe had a negative mass. In other words, there was some unknown force, essentially an anti-gravity, that was pushing space apart — and it is getting faster and faster.

Mauna Kea Observatories in Hawaii.

DARK ENERGY

Cerro Tololo observatory in Chile.

After going public a few months before the turn of the millennium, the lead researchers Saul Perlmutter, Brian Schmidt and Adam Riess tried to explain what they had found. The new form of energy was called dark energy. It was calculated that three-quarters of the energy in the Universe was made up of this dark mysterious phenomenon. Dark matter made up a fifth and that meant that all the stars and galaxies we can see make up just a twentieth of the total.

Dark energy is thought to be associated with vacuum energy. It appears that even nothing, the vacuum of space, has a tiny amount of energy in it. As the Universe expands, there is more and more nothing out there — so much that all of its energy has accumulated into the dominant entity in the Universe today. The discovery of dark energy tells us that the Universe is open. The Big Crunch will never happen, and the Heat Death of the Universe is much closer than first thought — in the next 20 billion years or so. The final event

of the Universe is now called the Rip. Dark energy will continue to pull the Universe apart. Galaxies will spread until all stars are so far apart the sky is perpetually dark, then atoms and molecules will be ripped asunder creating an impossibly thin soup of primordial particles spread over an infinite space. Some versions of the Big Bang theory suggest such a state of ultra-low energy was the starting condition for our Big Bang. In which case, has this all happened before?

The expansion of the universe is getting faster and faster.

The most recent history of science has involved discoveries on a huge scale carried out by immense experiments operated by thousands of scientists. These Herculean efforts have borne fruit that have set the direction for the next century of science. In astronomy, a new way of imaging the Universe has been discovered that maps gravity not radiation, and that looks set to allow us to see into the darkest recesses of space and time. In physics, the latest piece of

MASSIVE PROBLEM

The idea that some objects are heavier than others is an obvious and ancient one. Heaviness, or weight, is the force of gravity pulling on a mass, and it pulls harder on objects with more mass. A bucket of water is heavier than a single raindrop, but the material in both is identical. The raindrop is light purely because it is so much smaller than the bucket. Now imagine comparing a bucket

The ATLAS detector at the Large Hadron Collider was built to find the Higgs boson.

the Standard Model has been put in place with the discovery of the Higgs boson, a particle that solves one of the most thorny problems in physics, and could help to solve other mysteries.

full of water with a bucketful of mercury. They are the same size, but the liquid metal weighs 14 times as much. There must be 14 times as much mass in the same volume.

All this basic physics was settled in the early 19th century, and understood intuitively for centuries before. Mass was a measure of how much stuff was packed into materials. It was assumed that mercury atoms were simply bigger than the hydrogen and oxygen ones in water molecules. However, then particle physics showed that atoms were not solid lumps of matter at all. In fact they were largely empty, and their mass came almost entirely from the minute nucleus of protons and neutrons. (If an atom were the size of a football stadium, the electrons whizz around the stands and the nucleus would be the ball on the centre spot. The rest is empty!)

The mass of protons and neutrons, the most common type of a family of particles called hadrons, got their mass from the quarks inside them. A quark is the same size as an electron – in other words it is very small and takes up a single point in space (if that). However, a quark is 1,800 times more massive than an electron, and in the cases of highly exotic quarks, that figure leaps to billions of times heavier. And why do photons weigh nothing at all? No one could explain why.

2010 The Burj Khalifa in Dubai is completed and is currently the tallest building in the world at 829.8 m (2,2722.4 ft).
• *Solar Impulse* is the first aircraft to make a non-stop 24-hour flight using only solar power.

2011 The MESSENGER probe is the first spacecraft to go into orbit around Mercury.
• Nevada becomes the first state to allow self-driving cars on public roads.

2012 A survey estimates that the number of planets in the Universe is higher than the number of stars.
• The Large Hadron Collider finds evidence of the Higgs boson, a particle that gives matter its mass.
• The *Curiosity Rover* arrives on Mars and is lowered to the surface using rocket-power. It is tasked with exploring the Gale Crater.
• The Gansu Wind Farm in China is the largest wind-generating power plant.
• Xeno nuclei acid (XNA), an artificial form of DNA, is developed. This raises the possibility of an entirely new form of life being created in the laboratory.

2013 A 10-ton meteor explodes in the atmosphere above Chelyabinsk, Russia. The shockwave from the explosion causes windows to shatter and injures 1,200 people.

2014 The *Rosetta* spacecraft goes into orbit around Comet 67P/Churyumov–Gerasimenko and drops a lander to its surface. The spacecraft tracks the progress of the comet, watching how it changes as it swings around the Sun.

2015 The L0 maglev train in Japan reaches 965 k/h (600 mph) to set a new train speed record.
• The *New Horizons* probe makes a flyby of Pluto and Charon revealing them to be worlds made of nitrogen ice.

2016 LIGO, or the Laser Interferometer Gravitational-Wave Observatory, detects gravitational waves coming from two orbiting black holes. Gravity waves are hoped to offer a new way of observing the Universe.

2018 The Noor II in Morocco becomes the world's largest solar power plant.

HIGGS BOSON

Then in 1964 the British professor Peter Higgs and a few others had a go. They said that Universe was filled with a quantum field – now named after Higgs – which gave particles their mass. Without the field every particle, quark, electron, and photon were massless. Photons moved through the field without creating a disturbance, and so were unimpeded by the property of mass. Electrons create only small disturbances in the field, and have correspondingly small mass, while quarks create large disturbances. A disturbance makes the Higgs field bunch up around a particle, and this bunching is the Higgs boson, a waveform (or particle) that dictates how easily the particle moves through the field. The bosons impede the progress of quarks a lot and this translates into their high mass, and that creates all the stuff in the Universe that we can see, feel, and shift around.

It took the largest machine ever built to find the Higgs boson, the aptly named Large Hadron Collider (LHC). The LHC is run by CERN, the world's leading particle physics research centre, and is a 27-km (17-mile) subterranean ring located

Peter Higgs.

The LHC carries a beam of protons moving at just below the speed of light.

on the Swiss-French border. Around 1,600 superconducting magnets as big as trucks are used to focus protons into two beams and accelerate them in opposite directions to a tiny fraction of a per cent under the speed of light. The protons can never reach the speed of light, but as they inch closer their increase in energy gives them a greater momentum, which can be viewed as an increase in mass (although that is a simplified view). At the top speed, the protons in the LHC "weigh" 7,000 times more than when they started, and when two beams smash together they reach temperatures 100,000 times higher than at the core of the Sun. At this energy the protons form a "quake-gluon plasma", a primordial form of matter last seen in the Universe in the first second of its existence. It was hoped that the collisions would shows the Higgs field in action and in 2012, after 24 years in development, the LHC found the results it was built for.

The Higgs boson is still poorly understood, and the LHC, recently upgraded, is gearing up for further experiments which will help to answer questions such as why the Universe started out with more matter than antimatter, and whether dark matter is formed by mysterious super particles that shadow the known particles.

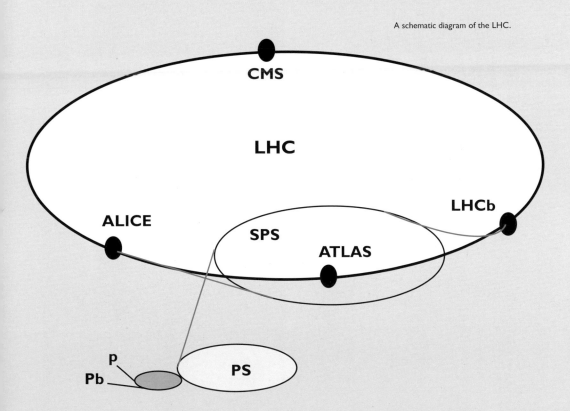

A schematic diagram of the LHC.

RIPPLES IN SPACE

Science never stops and the huge success of the Higgs boson discovery was soon old news. In 2016 the big story was the discovery of gravitational waves. Gravitational waves had been produced 100 years before by Albert Einstein as part of his general theory of relativity. This theory explained the force of gravity as a disturbance in the gravitational field caused by mass. These disturbances were in the very fabric of space and time, Einstein explained, and they rippled outward from moving objects just like the wash of a boat moving through water.

However, measuring gravitational waves proved very hard because they stretched and squeezed space itself, and everything in it including the wave detector. In addition, the actual changes were very small, and it took a global collaboration of scientists and laser arrays 4 km (2.5 miles) long to finally spot them.

The experiment was called LIGO, short for Laser Interferometer Gravitational-Wave Observatory. The LIGO detector used a powerful laser source that split into two beams which were sent off at right angles down a 4-km (2.5-mile) tube to a mirror at the far end. They bounced back and the two beams met again at the far end. One of the tubes was adjusted by a few billionths of a meter so the laser light had to travel a total of half a wavelength longer than the other tube. When these two beams met, their waves cancelled each other out completely.

The idea was that if the length of one tube was altered slightly by a gravitational wave, then the distance the laser travelled would also change. That change would mean that when the beams met they would no longer cancel out but leave a tell-tall flicker of light. That would be evidence of a gravitational wave. In reality, Earth's surface is trembling ever so slightly all the time due to volcanic activity deep underground. So LIGO built two sets of detectors, one in in Louisiana and the other in Washington State. Both these sites would experience a different pattern of vibrations from Earth and so the team could filter out those noise signals and search for the effects picked up by both detectors at the same time.

First cosmic event observed in gravitational waves and light.

BLACK HOLES BASH

On September 14, 2015, the gravity waves from a collision between two black holes a billion years ago reached Earth, and LIGO detected the tiny ripples left over from this monster event. Since then LIGO and a similar experiment running in Europe have been picking up more gravitational waves. At the moment they can only see tags that make an almighty splash in the cosmos, like neutron stars and black holes, but plans are afoot to build a LIGO system in space, which would send lasers between spacecraft in orbit around the Sun. The laser track would be 3,000,000 km (1,864,110 miles) long, dwarfing that of LIGO, and making it sensitive to a much wider spectrum of gravitation effects.

This space project, known as eLISA, is likely to take decades to develop, but being able to map the gravity of the Universe

Black holes colliding.

would be a game-changer for astronomy. It would allow us to look inside black holes, where no light ever shines, and see beyond the cosmic microwave background into the early days of the Universe. What we will find there could shift our view of the Universe entirely.

FUTURE OBSERVATIONS

The search is now on for more gravitational waves, and the best place to do it is away from the rumblings of Earth. In December 2015, LISA Pathfinder was launched. It is headed to an orbit at L1, a gravitationally stable position between the Sun and the Earth. There the spacecraft will test laser interferometry instruments in space, in the hope they can be deployed in a more ambitious experiment called eLISA. Provisionally scheduled for 2034, eLISA, or Evolved Laser Interferometer Space Antenna, will use three spacecraft triangulated around the Sun. Lasers will be fired between the spacecraft, making a laser track 3,000,000 km (1,864,110 miles) long, thus exponentially more sensitive to gravitational waves than LIGO.

The discovery of gravitational waves is a modern equivalent of Galileo's

The gravitational wave detector eLISA will be nearly a million times larger than the earthbound LIGO system.

observations of Jupiter's moons or Hubble's discovery of the expanding Universe, and gravitational astronomy looks set to change our view of the Universe forever.

No scientist thinks they will be out of a job very soon. In fact, science should continue forever. It will be impossible to know everything. If it does turn out that the Universe can be understood in its entirety, then that would appear to contradict everything we know about the Universe at the most fundamental level. If we did one day know everything, we'd have to go back and start again, because somewhere we'd made a mistake.

Science does not actually reveal certain facts, it just eradicates non-facts, and in so doing it creates a model of the way the Universe and everything in it works. As we progress with the scientific project, that model gets closer and closer to what we observe. One day the model may be so closely aligned with reality that we just give up trying to improve it. But we are not there yet. Let's see what the future of science has in its store.

WEBB TELESCOPE

Originally slated for launch in 2018, the James Webb telescope has been delayed due to technical issues. When it is launched, hopefully in the early 2020s, it will be the largest space telescope in history, and supersede the viewing power of the Hubble space telescope. Unlike Hubble, the Webb telescope images the universe in infrared, and that means its detector needs to keep very cold so it can detect the faint rays of heat from deepest space. The telescope has a 9-m (29.5-ft) gold-plated mirror for collecting the heat signals, all while shrouded by a heat shield

bigger than a tennis court. The spacecraft will be placed in orbit further out than Earth, where our planet's shadow will shield it from the heat of the Sun. The Webb will peer into dark nebulae to watch the development of protostars that are too young to emit light. It will allow us to see closer to the CMB, the edge of the visible Universe. This is where the oldest galaxies are and the first stars to form after the Big Bang. Out there, space is so stretched that the starlight has been redshifted into invisible heat. The Webb telescope will let us see these stars for the first time.

The Webb Telescope.

EXTREMELY LARGE

The European Extremely Large Telescope needs no further description other than it is being built in the mountain deserts of Chile, not Europe. Operated by the European Southern Observatory, the E-ELT will begin work in the mid 2020s. Its primary mirror will be 39.3 m (110 ft) wide, dwarfing Hubble's 2.4 m (7.9 ft) mirror. However, the E-ELT will have to look through the shimmering atmosphere which makes stars blur and twinkle. To solve this, light captured by the main mirror will be focused on to another with a very flexible surface. Underneath, 8,000 little pistons are wrinkling and warping the mirror 1,000 times a second to counteract the distortions of the atmosphere. This amazing telescope will be 15 times more powerful than Hubble, and that will allow it to resolve the exoplanets that orbit other worlds. That view will allow astronomers to see the absorption spectra of those alien worlds, and that spread of light will show what chemicals are in each planet's atmosphere. If the E-ELT is turned towards an Earth-like planet, a rocky world that is the right temperature for liquid water to exist on the surface, will it pick up the chemical signatures of life? If so, what then?

European Extremely Large Telescope.

SEARCHING FOR ALIENS

Astrobiology is a branch of life science that is looking for life beyond our world. If an astrobiologist is ever successful is finding aliens, then he or she will immediately shift disciplines and become an exobiologist. Exobiology is the study of alien life – only we've not found any yet. The number of exoplanets is estimated to be greater than the number of stars, so there is every chance the there are several billions of Earth-like planets in our galaxy alone. If the E-ELT or another project does find conclusive evidence of a biosphere around one of these worlds there is little chance of studying that world's life. There is a possible Earth-like exoplanet orbiting the Solar System's nearest stellar neighbour, Proxima Centauri. That star is 4 light years away. It would take centuries for a team of biologists to fly there aboard our current spacecraft technology. Even a radio signal to an alien civilization of Centaurians would take 8 years to receive a reply. The chance of alien civilizations matching our own is thought to be very low, but astrobiologists believe that simple single-celled life forms would be a relatively common phenomenon. In fact they believe that such alien life exists in our own Solar System in places like Jupiter's moon Europa and Saturn's Enceladus, which both have vast oceans of liquid water under their frozen surfaces.

Europa.

ARTIFICIAL INTELLIGENCE

It may appear limited and brutal at times, but human civilization was built on an evolved intelligence that must be very rare in the Universe. Although it is just one of billions of planets in an unspectacular corner of an average galaxy, Earth is a very special place. It is special because of its unusual Moon, heavy metallic core, stable star and particular set of neighbours, which add up into a world that has had the time to evolve complex life forms like humans.

However, it may be that the first scientists to travel the great distances needed to explore alien worlds will not be human but beings that use an artificial intelligence (AI). AI technology has been a focus of cutting-edge engineering for the last few decades. The fruits of this focus have not been quite what the sci-fi books told they would be.

An AI can be many things. Often it is an expert system, a vast database that has been filled with knowledge on a subject by human experts. This kind of AI then consults this data to make decisions. Often the expert system will work alongside another kind of AI technology called machine learning. This is where a computer programmes itself through trial and error so it can carry out an appointed task. It undergoes a period of training, running millions of iterations until the programme is perfected. It then uses that programme to analyze data – such as a picture, or a sound – and decide what it is and what to do with it (that where the expert system comes in).

This kind of AI is behind the voice recognition in smart speakers, search engines and social networks suggesting links that it thinks you want to see (as well as adverts), and pattern recognition used in medical and military applications. This kind of AI can do these jobs better and faster than we can, but that is all they can do. What about AI that is as clever us?

Smart AI systems can be programmed to explore the Universe.

THE INTERNET OF THINGS

It may be that what AI lacks is enough data. Some psychologists propose that the human consciousness and imagination which underwrites our innovating intelligence is a phenomenon that emerges from a subconscious that has reached a threshold of information. Our brains contain so much information that we have become conscious in order to manage it effectively. Would the same thing happen with machine intelligence? (Has it already happened?)

There are already many more devices connected to the Internet than there are people on Earth, and the number of devices is also growing very fast. These devices are not just computers and phones, they are washing machines, traffic lights, tills in shops, and weather buoys far out to sea.

This idea is called the Internet of Things, or IoT. All the items on the IoT produce a huge amount of data. Each piece is useful in its own way, but when combined in Big Data, it can transcend that use and be mined for information that we did not know was there. Who knows what we will discover.

QUANTUM COMPUTING

The requirements of Big Data and AI may be beyond the limits of silicon-based computing. There is a limit to how much smaller electronic components can get, which is fast approaching. There is a theoretical alternative that uses quantum characteristics for the switches in a computer's logic circuits. A classical computer's components are binary and so can only ever be on (1) or off (0). That means it holds just one bit of data (a 1 or a 0). The equivalent quantum component has a particular chance of being on or off and so is effectively on and off at the same time. Therefore, it holds two bits of information, not one. Things get more complicated very fast. A 32-bit classical computer handles 32 bits of information at once. A 32-quantum bit, or qubit, device can handle 4,294,967,296 bits in one go! That vast difference means that quantum computing could be able to tackle mathematics that is beyond current computing. However, the technology required to control atoms or subatomic particles to the extent that we can use them as a computer processor is still tantalizingly beyond our reach. If it is possible, science will find a way.

A prototype quantum computer.

SI UNITS

The SI Units

Base quantity	Name	Symbol
Length	meter	m
Mass	kilogram	kg
Time	second	s
Electrical Current	ampere	A
Thermodynamic temperature	kelvin	K
Amount of substance	mole	mol
Luminous intensity	candela	cd

These seven units are the basis of all measurement.

Measured quantity	Name of unit	Symbol	Definitions in base units	Alternative in derived units
Energy	joule	J	$m^2\,kg\,s^{-2}$	$N\,m$
Force	newton	N	$m\,kg\,s^{-2}$	$J\,m^{-1}$
Pressure	pascal	Pa	$kg\,m^{-1}\,s^{-2}$	$N\,m^{-2}$
Power	watt	W	$m^2\,kg\,s^{-3}$	$J\,s^{-1}$
Electric charge	coulomb	C	As	$J\,V^{-1}$
Electric potential difference	volt	V	$m^2\,kg\,A^{-1}\,s^{-3}$	$J\,C^{-1}$
Electric resistance	ohm	Ω	$m^2\,kg\,A^{-2}\,s^{-3}$	$V\,A^{-1}$
Electric conductance	siemens	S	$s^3\,A^2\,kg^{-1}\,m^{-2}$	$A\,V^{-1}$ or Ω^{-1}
Electric capacitance	farad	F	$s^4\,A^2\,kg^{-1}\,m^{-2}$	$C\,V^{-1}$
Luminous flux	lumen	lm	$cd\,sr$	
Illumination	lux	lx	$cd\,sr\,m^{-2}$	$lm\,m^{-2}$
Frequency	hertz	Hz	s^{-1}	
Radioactivity	becquerel	Bq	s^{-1}	
Enzyme activity	katal	kat	$mol\ substrate\ s^{-1}$	

DERIVED SI UNITS

Metric prefixes are pretty easy to understand and very handy for metric conversions. You don't have to know the nature of a unit to convert, for example, from kilo unit to mega-unit. All metric prefixes are powers of 10. The most commonly used prefixes are highlighted in the table.

Prefix	Symbol	Factor
yotta	Y	$10^{24} = 1,000,000,000,000,000,000,000,000$
zetta	Z	$10^{21} = 1,000,000,000,000,000,000,000$
exa	E	$10^{18} = 1,000,000,000,000,000,000$
peta	P	$10^{15} = 1,000,000,000,000,000$
tera	T	$10^{12} = 1,000,000,000,000$
giga	**G**	$10^{9} = 1,000,000,000$
mega	**M**	$10^{6} = 1,000,000$
kilo	**k**	$10^{3} = 1,000$
hecto	h	$10^{2} = 100$
deka	da	$10^{1} = 10$
deci	d	$10^{-1} = 0.1$
centi	c	$10^{-2} = 0.01$
milli	m	$10^{-3} = 0.001$
micro	μ	$10^{-6} = 0.000,001$
nano	n	$10^{-9} = 0.000,000,001$
pico	p	$10^{-12} = 0.000,000,000,001$
femto	f	$10^{-15} = 0.000,000,000,000,001$
atto	a	$10^{-18} = 0.000,000,000,000,000,001$
zepto	z	$10^{-21} = 0.000,000,000,000,000,000,001$
yocto	y	$10^{-24} = 0.000,000,000,000,000,000,000,001$

MAGNITUDES Ever wondered how many femtoseconds are in a minute or how big a petawatt is? All is revealed here. Most people even in the countries where the metric system is used only know the most important metric prefixes like 'kilo' and 'milli'. They are very handy for understanding metric conversions. The prefixes like 'zepto' or 'yotta' are very specific and used mostly in science.

BIBLIOGRAPHY

BOOKS:

Ponderables:
The Elements
Physics: An Illustrated History of the Foundations of Science
The Universe: An Illustrated History of Astronomy
Biology: An Illustrated History of Life Science
Engineering: An Illustrated History from Ancient Craft to Modern Technology

WEBSITES

The Scale of the Universe: http://htwins.net/scale2/
Tree of Life: http://tolweb.org/tree/
Dynamic periodic table: https://www.ptable.com/
The Standard Model: https://physics.info/standard/
Evolution game: https://keiwan.itch.io/evolution

YOUTUBE

Channel: Fermilab | Subject: Physics
Channel: PBS Space Time | Subject: Astronomy
Channel: Veritasium | Subject : Popular science

APPS

The Particle Adventure, Berkeley Lab
Galaxy Collider, Angisoft
Periodic Table, Royal Society of Chemistry

alloy: A mixture of metals.

alternating current: An electric current where the direction of flow changes at regular intervals.

alpha particle: A particle made of two protons and two neutrons, and with a charge of +2 that is produced by radioactive decays.

anion: A negatively charged ion.

anode: A positively charged electrode, which attracts negatively charged particles.

atom: The smallest unit of an element.

atomic number: The number of protons in an atom. The atoms of an element all have the same atomic number.

atomic mass: The average number of particles in the nucleus of an atom of a particular element.

catalyst: A substance that speeds up the rate of a chemical reaction but is not used up in the reaction.

cathode: A negatively charged electrode, which attracts positively charged particles.

cation: A positively charged ion; most cations are metallic.

charge: An electromagnetic property of subatomic particles; objects with opposite charges attract each other, while those with similar ones repel.

climate change: The gradual warming of Earth due to heat being trapped by gases in the atmosphere. Human activity, such as the burning of coal and oil, adds gases such as carbon dioxide to the air and exacerbates global warming.

compound: A substance made from two or more different elements bonding together.

conductor: A material that can carry heat or electricity well.

direct current: An electric current that flows in a constant direction.

electrode: A conductor that carries electric current into or away from an electric circuit.

electromagnet: A device that can have its magnetism turned on and off. It is generally an iron core surrounded by a coil of copper wire that temporarily generates a magnetic field when an electric current flows through it.

electromagnetic radiation: Oscillation in the electric and magnetic fields. Electromagnetic waves travel at the speed of light and include radio waves, microwaves, infrared, visible light, ultraviolet, and X-rays.

electron: A negatively charged particle found in atoms.

fission: When an atom splits into two roughly equal parts.

fusion: When two small atoms fuse to make a larger one.

ion: A charged particle created when an atom or molecule loses or gains an electron.

induction: When a magnetic field creates an electric current in a moving wire.

isotope: A version of an atom that has a different number of neutrons in its nucleus. All elements have several isotopes.

GLOSSARY

magnetism: Phenomena associated with magnets and magnetic fields. Magnetic fields are regions around magnets in which a force acts on any magnet or electric charge present.

molecule: The smallest part of a compound.

neutron: A component of an atom's nucleus. Neutrons have no electric charge.

neucleus: The core of an atom which contains almost all of the mass.

orbital: The space around an atom's nucleus filled with electrons.

proton: A positively charged particle that is located in the nucleus of every atom. Every element has a unique number of protons.

quark: A subatomic particle that makes up protons and neutrons, plus many and more exotic particles.

radiation: Giving off radioactive particles, heat, or electromagnetic waves.

radioactivity: A property of unstable elements, where the forces holding the atomic nucleus together are not able to hold so it eventually decays, releasing particles and energy.

resistance: A measure of how easily electric currents flow through a substance. Materials with a high resistance are insulators. A material with a low resistance is a conductor.

smelting: A chemical process used to purify metal.

superconductor: A substance that offers no resistance to an electrical current.

thermal: Relating to heat.

valence: A description of how many bonds an atom can form with other atoms.

voltage: The force that is pushing an electric current.

wavelength: The distance between the peak of a wave and the peak behind; the wavelength of a light wave is a measure of how much energy it contains.

INDEX

INDEX